10

Crystal Research and Technology

Journal of Experimental and Industrial Crystallography

Zeitschrift für experimentelle und technische Kristallographie

Volume 22
1987
Number 10 October

THEORIA CUM PRAXI

Akademie-Verlag · Berlin

ISSN 0232-1300 Cryst. Res. Technol., Berlin 22 (1987) 10, 1223–1340, K159–K172

Directions to the author

In the journal "Crystal Research and Technology – Kristall und Technik" as a rule the following categories of papers will be printed: original papers, review articles, short notes, and book reviews, preferentially in **English**. Complete information necessary for type-setting and a "Short Title" should be added.

The manuscript should not exceed 10 pages and be type-written on one side only in the format A4 (21 × 29.7 cm) on white ink-proof paper, with a two-lined distance and a left-hand free edge of 3 cm, and submitted in 2 copies. All sheets of the paper belonging to the contribution are continuously to be numbered.

The head of the manuscript should be arranged in this manner: initials and name of the author; institution, and title of the paper. Then two abstracts should follow, one in the original language and the second in German, English, French, or Russian, respectively.

Formulae, signs and symbols, letters etc. being wanting on the type-writer, should be distinctly written with penholder. O as zero is to be underlined, the capital O is not. Greek letters should be underlined with red ink.

Paragraphs should be marked by indenting of at least 3 touches.

Concerning the marking of the manuscript should be noted:
headings should be underlined, sub-headings not, and indented by 9 touches. The titles of tables are not to be underlined. Other markings in the text will be done by the editorial staff. They mean:

light-red ～～～～～～	= small capitals	light-red: perpendicular straight red line on the left edge = small printing
blue ～～～～～～	= in italics	

Tables are to be added on separate sheets of paper at the end of the contribution.

Foot notes should be continuously written on a separate sheet of paper.

Figures (Fig.) – drawings and photographs – should be not more than 8 and submitted (2 copies) in a special envelope: photographs as white high-gloss prints (inscription on a transparent covering sheet) and drawings as distinct sketches. On the back of the submitted figures the reduction scale in tenth (9/10, 8/10, 5/10 e.g.), author's name, short title of the paper, and the number of the figure should be given.

Captions for the figures should be submitted on a separate sheet of paper.

References should be written without numbers in alphabetical order of the names at the end of the contribution. Cited names in the text stand for the references. The references should contain the following items, for example:

for books: RAST, K.: Physikalisch-chemische Rechnungen, Berlin 1958
for journals: KRAUSE, H. N., MÜLLER, K.: Cryst. Res. Technol. **16**, 70 (1981)

Add, please, the short title, about 6 keywords, and the academic grades of the authors.
The authors will be sent two galley proofs of review articles and original papers. Proof reading of short notes and book reviews will be done by the editorial staff.

For each of these papers a total of 50 reprints is available free of charge. Requests for additional maximum 200 reprints (against payment) should be given together with the submitted manuscript.

Founded 1966 as
Kristall und Technik

Crystal Research and Technology

10

Journal of Experimental and Industrial Crystallography

VOLUME 22 · 1987 · NUMBER 10

Akademie-Verlag · Berlin

Papers should be directed to the editor's office
DDR-7030 Leipzig, Scharnhorststraße 20,
or to the editor Prof. Dr. Hermann Neels,
DDR-7254 Machern, Zeititzer Weg 25,
or to one of the members of the Advisory Board.

Orders can be sent

- in the GDR:
  to the Postzeitungsvertrieb or to the Akademie-Verlag Berlin, Leipziger Straße 3–4, DDR-1086 Berlin;
- in the other socialist countries: to a book-shop for foreign language literature or to the competent news-distributing agency;
- in the FRG and Berlin (West): to a book-shop or to the wholesale distributing agency Kunst und Wissen, Erich Bieber OHG, Wilhelmstr. 4–6, D-7000 Stuttgart 1;
- in the other Western European countries: to Kunst und Wissen, Erich Bieber GmbH, General Wille-Str. 4, CH-8002 Zürich;
- in USA and Canada to the VCH Publishers, Inc., 303 NW 12th Avenue, Deerfield Beach, FL 33442-1788, USA;
- in other countries; to the international book- and journal-selling trade, to Buchexport, Volkseigener Außenhandelsbetrieb der Postfach 160, DDR, DDR-7010 Leipzig, or to the Akademie-Verlag Berlin, Leipziger Straße 3–4, DDR-1086 Berlin.

Journal „Crystal Research and Technology"

Herausgeber: Im Auftrage des Akademie-Verlages Berlin herausgegeben von Prof. Dr. Hermann Neels.

Verlag: Akademie-Verlag Berlin, Leipziger Str. 3–4, DDR-1086 Berlin; Fernruf: 2236221 oder 2236229; Telex-Nr.: 114420; Bank: Staatsbank der DDR, Berlin, Kto.-Nr. 6836-26-20712.

Chefredakteur: Prof. Dr. Hermann Neels, Zeititzer Weg 25, DDR-7254 Machern; Fernruf: Brandis 3442. Anschrift der Redaktion: Scharnhorststr. 20, DDR-7030 Leipzig; Telefon: 310502.

Veröffentlicht unter der Lizenznummer 1526 des Presseamtes beim Vorsitzenden des Ministerrates der Deutschen Demokratischen Republik.

Gesamtherstellung: VEB Druckerei „Thomas Müntzer", DDR-5820 Bad Langensalza.

Erscheinungsweise: Das Journal „Crystal Research and Technology" erscheint monatlich. Die 12 Hefte eines Jahrganges bilden einen Band. Bezugspreis je Band 480,– DM zuzüglich Versandspesen; Preis je Heft 40,– DM. Der gültige Jahresbezugspreis für die DDR ist der Postzeitungsliste zu entnehmen.

Bestellnummer dieses Heftes: 1082/22/10

AN (EDV) 35830
02800

Cryst. Res. Technol.	22	1987	10	1225–1228

Original Papers

Miloš Nevřiva

Institute of Physics, Czechoslovak. Academy of Sciences
Prague, Czechoslovakia

Crystal Growth of Mg- and Ti-substituted Hexagonal Ferrites from PbO-B_2O_3 Based Melts

Crystals of Mg and Ti-substituted hexaferrites have been prepared as perspective materials for the photoelectrochemical conversion of solar energy by means of anodes. The melts on the basis of PbO–B_2O_3 were used to grow both bulk crystals and thin films. In the chosen system the primary crystallization field of hexaferrites was found. The crystals were grown by conventional unseeded crystal growth using (ACRT) accelerated crucible rotation technique to ensure their good quality and homogeneity. The size of the crystals varied from 10 to 23 mm in diameter. Maximum total content of Mg^{2+} and Ti^{4+} amounted to 2 atoms per formula unit of hexaferrite.

Einkristalle der Hexaferrite wurden im Rahmen von Forschungen zu neuen Materialien für die photoelektrische Umwandlung der Sonnenenergie vorbereitet. Hinsichtlich der Möglichkeit der Herstellung dünner Schichten der genannten Materialien wurde eine aus PbO und B_2O_3 bestehende Schmelze gewählt. Im gewählten System wurde das Gebiet der primären Kristallisation der Hexaferrite gefunden. Die Einkristalle wurden durch die klassische Methode der Kristallzüchtung aus fremder Schmelze unter Ausnutzung der gesteuerten Keimbildung durch die Kühlung eines Punktes auf dem Boden des Tiegels vorbereitet. Die Homogenität der Einkristalle wurde durch das verlangsamte und beschleunigte Drehen des Tiegels (ACRT Methode) erreicht. Die gezüchteten Einkristalle waren bis zu 25 mm groß. Die Experimente zeigten, daß man im gewählten System die Einkristalle mit einem Gehalt an Mg^{2+} und Ti^{4+} bis zu 2 Atomen pro Formeleinheit des Hexaferrits züchten kann.

1. Introduction

For the past several years a great interest was devoted to the chemical conversion and storage of solar energy by means of photoelectrodes. Hexagonal ferrites appear to be one class of the perspective materials for this purpose (Pollert et al. 1983, 1986). Their photoelectrolytic properties particularly electron affinity can be improved by the substitution of Mg^{2+}–Ti^{4+} pairs for iron ions (Hejtmánek et al.). To increase the protection of these materials for photoelectrodes against photocorrosion in an aqueous solution Pb ions were replaced by Ba. Single crystals of $Pb_xBa_{1-x}Mg_yTi_zFe_{12-y-z}O_{19}$ were grown from PbO–B_2O_3 based melts with the aim of growing both bulk crystals and thin films. The thin films have not been prepared yet due to the lack of suitable substrates. In the chosen system the primary crystallization field of hexaferrites was determined.

2. Experimental

Single crystals have been prepared by a conventional unseeded crystal growth on the bottom of 100 cm³ platinum crucible using ACRT. The experiments were performed with 250 g of the melt in a vertical tube furnace with Kanthal heating element (Kanthal REH B 10/60). The temperature regime was controlled by a PID controller (KFKI Hungary) with an accuracy better than 0.2 K.

Each experimental run was started by the determination of the saturation temperature, T_S (see Table 1), taken as the temperature at which the first crystals appeared on the thermocouple Pt–Pt10Rh dipped into homogeneous melt. The appearing of the crystals was optically observed on the thermocouple pulled off from the melt after each decrease of the temperature by about 10 K at the rate of 0.2 K h^{-1}. The saturation temperature so can be determined in this way with an accuracy better than ± 5 K. To ensure the nucleation only on the thermocouple a temperature lower by 1 to 3 K at the surface of the melt was maintained. After the temperature was increased about 100 K above the saturation temperature and the temperature gradient at the crucible was changed to have a lower temperature at the bottom and crystal growth by spontaneous nucleation using cooling rate of 0.2 K h^{-1} was started. The number of nuclei created during the temperature decrease was minimized by a spot cooling of the bottom of the crucible by a platinum rod. The crucible rotation, the regime which had already been described (NEVŘIVA et al.), was started when the saturation temperature was reached. At the same time the rate of cooling was increased to 0.5 K h^{-1}.

Table 1

Survey of melt compositions and saturation temperatures

Run No.	R_1	R_2	R_3	R_4	R_5	R_6	T_S °C
1	0.0449	—	6.450	0.441	—	—	1190
2	0.0672	—	6.550	0.447	—	—	1185
3	0.0672	2.478	6.444	0.348	0.1100	0.1770	1174
4	0.0672	1.662	6.444	0.338	0.1440	0.2320	1202
5	0.0672	9.980	6.444	0.404	0.0348	0.0562	1178
6	0.0672	6.700	6.444	0.390	0.0754	0.0544	—
7	0.0672	10.370	6.450	0.403	0.0227	0.0652	1132
8	0.0672	5.775	6.440	0.393	0.0380	0.1096	1178
9	0.0000	10.370	6.436	0.380	0.0227	0.0652	1198

The chemicals used for the growth of $Pb_xBa_{1-x}Mg_yTi_zFe_{12-y-z}O_{19}$ crystals were chemically analysed and the correction coefficients for the calculation of the real melt composition were determined on the basis of this analysis.

The melt composition data for experimental runs are summarized in Table 1 where R-coefficients represent the following atomic ratios:

$$R_1 = \frac{BaO}{PbO + BaO}\,; \qquad R_2 = \frac{2\,Fe_2O_3}{TiO_2 + MgO}\,, \tag{1}$$

$$R_3 = \frac{PbO}{2\,B_2O_3}\,; \quad R_4 = \frac{2\,Fe_2O_3 + TiO_2 + MgO + BaO}{2\,Fe_2O_2 + TiO_2 + MgO + BaO + PbO + 2\,B_2O_3}\,, \tag{2}$$

$$R_5 = \frac{MgO}{MgO + TiO_2 + 2\,Fe_2O_3}\,; \qquad R_6 = \frac{TiO_2}{MgO + TiO_2 + 2\,Fe_2O_3}\,. \tag{3}$$

3. Results and discussion

The relationship between the composition of the melts and composition of the hexaferrite crystals can be described in the form of the following distribution coefficients K_i

$$K_{Mg} = \frac{(y/12)_{crystal}}{[MgO/(MgO + TiO_2 + 2\,Fe_2O_3)]_{melt}} = \frac{y}{12R_5}, \quad (4)$$

$$K_{Ba} = \frac{(x)_{crystal}}{[BaO/(BaO + PbO)]_{melt}} = \frac{x}{R_1}, \quad (5)$$

$$K_{Ti} = \frac{(z/12)_{crystal}}{[TiO_2/(MgO + TiO_2 + 2\,Fe_2O_3)]_{melt}} = \frac{z}{12R_6}, \quad (6)$$

where x, y, z express the number of atoms of the corresponding component in the chemical formula of hexaferrite.

The chemical composition of prepared crystals and the obtained distribution coefficients are listed in Table 2. As it can be seen, in the Tables 1 and 2, the content of Mg^{2+} and Ti^{4+} in hexaferrite crystals increased with decreasing value of R_2 coefficient. The highest total content of both ions was obtained when R_2 was equal to 5.775. An additional decrease of R_2 led to the primary crystallization of spinel crystals the size of which was about 10 mm in diameter. In contrast to that the size of hexaferrite crystals varied from 10 to 23 mm. The biggest one yielded by run No. 8 is shown in Figure 1. Inequality of the number of Mg and Ti ions being mostly found arose from different values of distribution coefficients of both components. As a consequence of this fact

Table 2

Crystal compositions and distribution coefficients

Run No.	crystal composition [number of atoms per formula unit]	K_{Ba}	K_{Mg}	K_{Ti}
1	$Pb_{0.20}Ba_{0.80}Fe_{12}O_{19}$	17.59	—	—
2	$Pb_{0.20}Ba_{0.80}Fe_{12}O_{19}$	11.90	—	—
3	spinel — $Mg_{1.04}Fe_{1.92}Ti_{0.04}O_4$	—	—	—
4	spinel — $Mg_{1.09}Fe_{1.80}Ti_{0.11}O_4$	—	—	—
5	$Pb_{0.32}Ba_{0.47}Mg_{1.06}Ti_{0.52}Fe_{10.89}O_{19.21}$	6.99	2.538	0.771
6	$Pb_{0.27}Ba_{0.63}Mg_{0.91}Ti_{0.45}Fe_{10.97}O_{19.10}$	9.37	1.006	0.689
7	$Pb_{0.31}Ba_{0.62}Mg_{0.50}Ti_{0.50}Fe_{11.08}O_{19.07}$	9.23	1.835	0.639
8	$Pb_{0.16}Ba_{0.73}Mg_{1.13}Ti_{0.69}Fe_{10.47}O_{19.11}$	10.86	2.478	0.525
9	hexaferrite — not analyzed	—	—	

Fig. 1

there exists an improper ratio of cations to oxygen anions in the lattice of hexaferrite together with a variation of valences of Fe ions.

To determine the homogeneity of the hexaferrite crystals with respect to the distribution of Ba ions, some crystals obtained at runs No. 1 and 2 were cut into parts and each of them was chemically analyzed separately. It was found that the content of Ba gradually decreased from the highest content of the crystal part corresponding to the initial stage of a crystal growth down to the lowest Ba content belonging to the crystal part growing at the end of the crystal growth (at the lowest temperature). The total decrease of the Ba content did not exceed 50 rel.% (in Table 2 average values are given). These variations are probably caused by the rapid exhaustion of Ba component from the melt due to much higher value of K_{Ba} than unity. Similar action, however smaller, can be also expected with the segregation of Mg.

Author wishes to thank RNDr. D. Zemanová for careful chemical analysis and Dr. E. Pollert for helpful discussion and suggestions.

References

POLLERT, E., HEJTMÁNEK, J., DOUMERC, J. P., CLAVERIE, J., HAGEMULLER, P.: J. Phys. Chem. Solids **44**, 273 (1983)
POLLERT, E., HEJTMÁNEK, J., DOUMERC, J. P.: Z. anor. allg. Chem. 540/541, 205 (1986)
HEJTMÁNEK, J., POLLERT, E., DOUMERC, J. P.: Coll. Abstr. of 6th Conf. on Photochemical Conversion and Storage of Solar Energy, D-49, Paris 1986
NEVŘIVA, M., ŠIMŠOVÁ, J., POLLERT, E., ČERMÁK, J.: J. Cryst. Growth **71**, 782 (1985)

(Received April 3, 1987)

Authors' address:

MILOŠ NEVŘIVA
Institute of Physics, Czechoslovak Academy of Sciences
18040 Prague 8
Na Slovance 2
Czechoslovakia

Cryst. Res. Technol. | 22 | 1987 | 10 | 1229–1234

M. SUSZYŃSKA, A. GUBAŃSKI, H. OPYRCHAŁ

Institute of Low Temperature and Structure Research, Polish Academy of Sciences
Wrocław, Poland

Low-temperature Precipitation in NaCl: Eu^{++} Monocrystals[1])

It has been shown that the macroscopic features of the as grown NaCl crystals containing up to about 140 m ppm of Eu^{++} ions agree well with the main characteristics of the Na_2EuCl_4 phase present in these crystals.

Thermal decomposition of the NaCl:Eu^{++} solid solution leads to the formation of the mentioned Na_2EuCl_4 phase at temperatures not very much higher than 273 K. The kinetics of this process is governed by the pipe diffusion which leads to the growth of the small impurity aggregates already present in quenched samples.

1. Introduction

The NaCl:Eu^{++} system was recently the subject of our systematic investigations. Among others, the concentration dependences of optical, dielectrical and mechanical characteristics of as grown (AG) samples were studied (SUSZYŃSKA et al.). On the other hand, the europium-related precipitates present in a 2-years-old crystal nominally doped with 0.1% of the dopant, have been studied by X-ray methods (STĘPIEŃ-DAMM, ŁUKASZEWICZ).

The aim of the reported work is twofold:
1. to compare the macro- and microscopic data for AG crystals which already contain the Na_2EuCl_4 phase, and
2. to provoke the precipitation of this phase in the solution treated (ST) samples by annealing at appropriate thermal conditions.

In this regard the present work substantially extends and supplements the previous results (SUSZYŃSKA et al.; STĘPIEŃ-DAMM, ŁUKASZEWICZ; LOPEZ et al.; MARTIN et al.; OROZCO et al.; CAPELLETTI, MANFREDI; MUGEŃSKI, CYWIŃSKI).

To make the problem of the data compatibility entirely clear, we shall reemphasize some of the informations concerning the Na_2EuCl_4 phase obtained by X-ray investigations (STĘPIEŃ-DAMM, ŁUKASZEWICZ):
1. Preferential localization near subgrain boundaries,
2. small thermal stability,
3. disappearance under the X-ray action used for the test performance,
4. the unit cell could be of CsCl-type, where the corners are occupied by cation vacancies and the europium as well as sodium cations; the lattice parameter is equal to 7.47 Å.

In the first part of this report it is assumed that the Na_2EuCl_4 phase is formed immediately after the solubility limit of the dopant becomes exceeded. A microstructural model of this phase, compatible with the available data, is postulated. The correctness of this hypothesis is evidenced by the results described in the second part of this work.

---

[1]) The results of this work have been partly presented as poster at the Fifth Europhysical Topical Conference "Lattice Defects in Ionic Crystals", Madrid 1986, and published as extended abstract in the Proceedings of this Conference: P. II.16, p. 225–226.

2. Na_2EuCl_4 — Phase in AG crystals

2.1. Some experimental details

If not otherwise stated, the measurements of ionic thermocurrent (ITC), optical absorption, photoluminescence and the stress-strain curves for the AG crystals have been performed according to the generally applied techniques, cp. see for instance Suszyńska et al.

The content of europium expressed in mole pp, is proportional to the value of the absorption coefficient of the high-energy band in the ST state, $\alpha(e_g)$, expressed in cm^{-1}. The proportionality factor (A = concentration/$\alpha(e_g)$) is equal to 4.7 for crystals with $\alpha(e_g) \lesssim 20\ cm^{-1}$. For higher dopant contents the quenching procedure used, i.e. the dropping of the hot sample onto a copper block at room temperature, did not prevent some aggregation already during the cooling period. For this reason instead of the dopant concentration the values of $\alpha(e_g)$ are considered throughout the work reported.

2.2. Structure-sensitive properties

2.2.1. Optical characteristics

Figure 1 presents some optical characteristics of the AG crystals in which the Na_2EuCl_4 phase should be the dominant one. The main experimental findings are as follows:

1. Absorption spectrum exhibits the smallest value of the $10D_q$-parameter (from among the four spectra typical of variously doped crystals) indicating the smallest splitting effect of the cubic crystal field at the related Eu^{++} sites.

2. Although the emission spectrum, taken at the liquid nitrogene temperature for the AG- and ST-samples, is peaking at 427 nm, its width is distinctly larger than in the case of isolated impurity-vacancy (I.V.) dipoles, solely present in the quenched, slightly doped samples.

3. Although the broad emission band may be characteristic of the europium ions in several dissimilar environments, the distribution of their transition probabilities cannot be very large because the luminescence decay-time curves observed remain exponential in the prevailing part of the time scale considered.

4. For AG samples the lifetime value (τ) is slightly temperature-dependent and practically independent of the wavelength of the light emitted. The value of τ is much shorter (615 ns) than the radiative lifetime characteristic of isolated I.V. dipoles (1050 ns) and of the loosely-bounded dimers (750 — 860 ns) as well, cp. Capelletti, Manfredi.

On the basis of the above presented data the emission spectrum of the crystals under consideration can be interpreted in two different ways: either precipitates or I.V. dipoles together with some small dipole aggregates are present in both crystal states being responsible for the main — 427 nm — band.

2.2.2. ITC-data

The ITC-measurements showed, among others, that in the crystals with relatively high density of dislocations, either grown-in or induced by plastic deformation, the dopant ions exhibit a pronounced tendency towards segregation near the dislocation lines. This Cottrell-like atmosphere yields a characteristic ITC-band sensitive to high-temperature annealing, plastic deformation and ionizing radiation (Suszyńska, Capelletti) resembling the behaviour of the Na_2EuCl_4 phase considered. The effect of plastic deformation is recoverable at temperatures near RT, and the process is governed by the $t^{2/3}_{\text{ageing}}$-law, suggesting the dopant segregation near the dislocations.

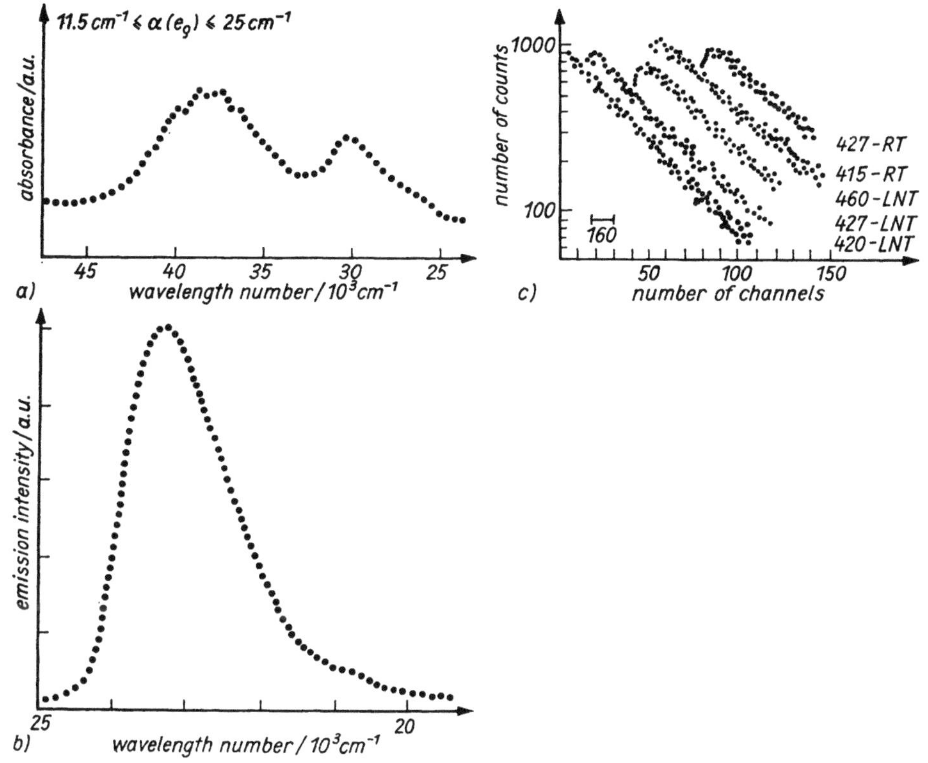

Fig. 1. Optical data concerning the NaCl crystal containing approximately 122 mole ppm of Eu^{++}: a) absorption at RT, b) emission at LNT, c) photoluminescence decay – time curves vs $\lambda_{em.}$ and T_{test}

The activation energy of the recovery process occurring for the AG-samples is equal to the activation energy for the migration of cation vacancies in NaCl crystals, indicating a vacancy-aided process. It seemed reasonable to assume the cation vacancies are chemical constituents of the Na_2EuCl_4 particles preferentially segregated near the dislocation lines.

2.2.3. Mechanical response

For the mechanical characteristics of the precipitation-strengthened materials important are the shape of the precipitated particles and the nature of the precipitate/matrix interface. Although the precise geometry of the Na_2EuCl_4-precipitate particles is unknown at present, the available informations are sufficient to indicate the Kelly/Nicholson-cutting process (KELLY, NICHOLSON) as the basic interaction mechanism responsible for the yield stress of the crystals under considerations. The most important fact which emerges from these data is the requirement of coherency of the Na_2EuCl_4 particles with the NaCl matrix which cannot be fulfilled by the bcc-type inclusion in the fcc-type host lattice.

This apparent discrepancy can be removed by assuming that the structural model is somewhat more complex than the originally postulated one (STĘPIEŃ-DAMM, ŁUKASZEWICZ). Figure 2 shows the mutual orientation (1 from among the 3 admissible ones) of NaCl and Na_2EuCl_4 unit cells. This ordered array of planar dimers matches

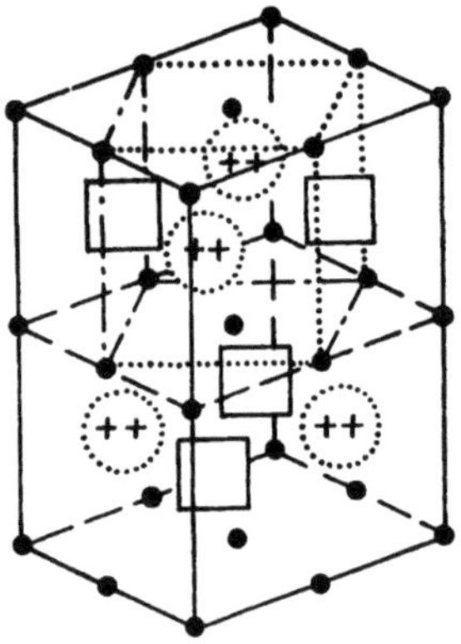

Fig. 2. Cation sublattice of the Na_2EuCl_4 particles: ○ mark the Na^+ and Eu^{++} ions, respectively, □ – the cation vacancies. Inner cube represents the NaCl unit cell

well the NaCl matrix, where the lattice misfit amounts to about 6% and is accomodated by the stress fields related with the dislocation lines.

3. Annealing experiments

The studies of low-temperature annealing of ST-crystals containing an amount of europium equivalent to $\alpha(e_g) \gtrsim 40\ cm^{-1}$ were undertaken in an attempt to obtain a detailed picture of the precipitation process, and to examine the role of dislocations in this process. Because plastic deformation promotes the precipitation in a heterogeneous regime, some of the ST-samples were subjected to the compression at RT; the strain rate was equal to $3 \times 10^{-4}\ s^{-1}$. Although the results obtained are of preliminary character on this time, several interesting features of the thermally induced processes have been monitored.

3.1. Effectiveness of the solution treatment

As it was mentioned above, the normal procedure for the solution heat treatment of the crystals (0.5 h at 873 K before quenching) does not prevent some aggregation already during the cooling period to RT. Comparing our ST-data with those obtained by CAPELLETTI, MANFREDI, the loosely-bound dimers are supposed to be the first products of aggregation. It was pertinent to consider what causes the stability of these aggregates and in what way these aggregates assist the nucleation and/or growth of the Na_2EuCl_4 precipitate particles. It should be taken into account, however, that the presence of quenched-in small aggregates makes a rigorous analysis of the precipitation kinetics difficult.

3.2. Precipitation kinetics

After solution treatment some of the samples were subjected to the room temperature storage while other were tested after a low-temperature annealing. The ageing was done isothermally for T_{ag} between RT and 333 K and t_{ag} between few minutes and few months.

The analysis of the precipitation kinetics which rules the decrease of I.V. dipoles, monitored either by ITC or $\alpha(e_g)$ at the wavelength value which corresponds to the absorption maximum in ST-state, shows that a double logarithmic plot of the reciprocal of the unprecipitated fraction of dipoles vs t_{ag} yields a straight line with the slope equal to 0.67, cp. figure 3, as expected for the Na_2EuCl_4 particles. The process

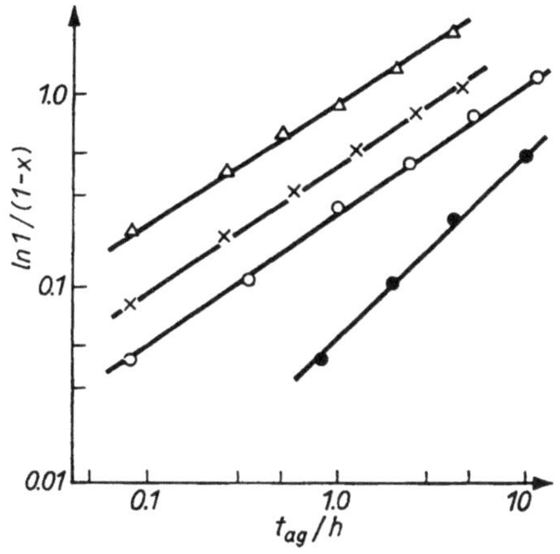

Fig. 3. Logarithm of the reciprocal of the unprecipitated fraction of dipoles vs the ageing time for a crystal with $\alpha(e_g)$ in ST-state equal to 60 cm^{-1}. ● 293 K, ○ 303 K, × 313 K, △ 323 K

is best described by a dislocation-induced precipitation (HARPER) due to the diffusion of dipoles and dimers towards the dislocation lines. It is reasonable to assume that the dislocations "contain" many jogs which could be either nucleation sites or short-circuit paths of the enhanced diffusion of the dopant ions. The determined value of the activation energy was equal to 0.70 eV suggesting that the growth of the already present nuclei of precipitates is the rate-controlling process rather than the nucleation itself.

3.3. Concluding remarks

Preliminary studies of the europium-related precipitation kinetics in NaCl crystals unambiguously showed that the best model for the process occurring just above room temperature is the motion of the dopant-related clusters towards the dislocation lines. Still more favourable are probably the low-angle grain boundaries which play the dominant role under the crystal-growth conditions.

On the basis of the results obtained it is possible to devise the following sequence for the process considered:

(I.V. dipoles + loosely bound dimers) in the ST-state → quadripoles which at RT undergo stabilization near the dislocation lines → ordering at $T_{ag} \gtrapprox$ RT due to the enhanced diffusion within the dislocation cores → Na_2EuCl_4 particles which — in heavily doped samples may coexist with the extraordinary $EuCl_2$-phase with the structure of CaF_2 as has been shown rentgenographically; for slightly doped crystals this phase can be induced thermally by annealing the ST-samples at $T_{ag} \gtrapprox 353$ K (SUSZYŃSKA).

Because the thermal history of samples, and especially the procedure of the solution treatment, is critically important in establishing the details of the process postulated, additional systematic studies are necessary, e.g. for crystals containing 100% of isolated dipoles. Also the catalytic role of vacancies and dislocations as well should be determined more precisely.

The authors are indebted to Dr. J. Stępień-Damm and Dr. M. Czapelski for many helpful discussions.

References

Capelletti, R., Manfredi, M.: phys. stat. sol. (a) **86**, 333 (1984)
Cottrell, A. H., Bilby, A. B.: Proc. Soc. A. **62**, 49 (1949)
Harper, S.: Phys. Rev. **83**, 709 (1951)
Kelly, A., Nicholson, R. B.: Progress in Material Sciences (B. Chalmers Oxford, New York, 1963, Vol. 10, p. 142
Lopez, F. J., Murrieta, H. S., Hernandez, J. A., Rubio, J. O.: Phys. Rev. B **22**, 6428 (1980)
Martin, J. J., Harness, V. J., Orozco, E. M., Mendoza, A. A., Rubio, J. O.: Phil. Mag. A **46**, 629 (1982)
Mugeński, E., Cywiński, R.: phys. stat. sol. (b) **128**, K75 (1984)
Orozco, E. M., Soullard, J., Zaldo, C., Agullo-Lopez, F.: Phil. Mag. **50**, 423 (1984)
Stępień-Damm, J., Łukaszewicz, K.: Acta Cryst. A **36**, 54 (1980)
Suszyńska, M.: to be published
Suszyńska, M., Capelletti, R.: Cryst. Res. and Technol. **19**, 1385, 1489 (1984); **20**, 1363 (1985)
Suszyńska, M., Opyrchał, H., Czapelski, M., Manfredi, M.: J. Mater. Sci., in print, 1986

(Received March 9, 1987)

Authors' address:

Dr. sc. Maria Suszyńska, A. Gubański, PhD., H. Opyrchał, PhD.
Institute of Low Temperature and Structure Research, Polish
Academy of Sciences
P.O. Box 937
PL 50950 Wrocław
Poland

Cryst. Res. Technol.	22	1987	10	1235–1240

O. SÖHNEL, J. W. MULLIN

University College London, Department of Chemical and Biochemical Engineering London, England

Influence of Mixing on Batch Precipitation

The influence of mixing on the induction period of batch precipitation of Ag_2WO_4, $CaWO_4$ and $SrWO_4$, in the supersaturation ranges 3.7–7.8, 32–134 and 121–210 respectively, has been studied at 25 °C. Increasing the intensity of mixing invariably decreases the induction period in all three systems. This effect in absolute terms is more pronounced at the lower supersaturations than at the higher.

1. Introduction

The effects of mixing on the primary (i.e. nucleation and crystal growth) and secondary (e.g. agglomeration, recrystallization, crystal breakage and attrition) processes during the crystallization of moderately soluble substances have long been recognized. From the industrial point of view, mixing often plays a decisive role in crystallizer performance and the physical form and composition of the crystalline product (MULLIN; JANČIĆ, GROOTSCHOLTEN; CHIANESE et al.).

In general, increasing the intensity of mixing brings about increases in the rates of nucleation and of diffusion-controlled crystal growth (NÝVLT et al., 1985) and a decrease in the apparent activation energy of crystallization (TAKUBO). These effects are undoubtedly responsible for decreasing both the induction period of crystallization (e.g. GARSIDE et al.) and the metastable zone width (e.g. NÝVLT et al., 1966).

In the crystallization of moderately soluble substances, a low level of supersaturation is normally generated relatively slowly, e.g. by the gradual cooling of a saturated solution, and no major supersaturation fluctuations are expected so long as the system is reasonably well-mixed. In a precipitation process, however, i.e. the fast crystallization of a sparingly soluble substance, the situation is quite different. A high level of supersaturation is usually established by rapidly mixing together two distinct liquid phases in an attempt to bring them into intimate contact at the molecular level.

In the case of precipitation, therefore, mixing would be expected to exert a substantial influence on the rate of development of supersaturation and its distribution within the system. Consequently, mixing might be expected to influence the mechanism and rate of both nucleation and crystal growth, the number of crystals formed, their size, shape and crystalline modification, crystal agglomeration and ageing, making the overall effects of mixing on the precipitation process exceedingly complex.

There is experimental evidence available that the method by which solutions are mixed together influences substantially the rate of precipitation and the size of the resulting crystals (TURNBULL; O'HERN, RUSH). The intensity of mixing has been shown to change the number of crystals produced and their size distribution (SÖHNEL 1981; POHORECKI, BALDYGA 1983, 1985), and the habit and crystalline form (LIU et al.; Roquez, Girou). It has also been reported that for moderately soluble substances, specifically nickel ammonium sulphate and potash alum, increasing the intensity of mixing reduces the induction period to a certain value beyond which no further significant effects can be observed (Mullin, Osman; Mullin, Žáček).

The present study aims to establish whether mixing exerts an influence on the induction period of batch precipitation of sparingly soluble substances, and if so, whether there are similarities to the behaviour of moderately and highly soluble substances. The possibility of inferring basic information about a precipitating system from induction period data is also explored. The substances chosen for precipitation were silver, calcium and strontium tungstates.

2. Experimental

50 mL of $AgNO_3$, $CaCl_2$ or $SrCl_2$ solution was rapidly poured into the same volume of equimolar (half-equimolar in the case of Ag_2WO_4 precipitation) solution of Na_2WO_4 agitated in a baffled beaker, all solutions were kept at 25 °C. Agitation was effected by a two-blade paddle-wheel glass stirrer, with adjustable speed, which rotated in the solution without any contact with the beaker bottom or the four wall baffles. Solutions were prepared by dissolving Analar grade chemicals in distilled water, passing through a 0.4 μm membrane filter and leaving for at least one week to age before use.

Induction periods were determined by visual observation of the first appearance of turbidity viewed against a black background. The time elapsed between solution-addition and turbidity-appearance was taken as the duration of the induction period. Experiments were carried out at agitator speeds of 300, 500 and 700 rpm respectively. Three independent determinations of induction period were made for each concentration and their arithmetic average was taken as the representative value for the given conditions.

3. Results

In all cases, increased mixing intensity, quantified here by agitator speed, resulted in a decrease in the induction period, as illustrated in Figure 1 for the precipitation of Ag_2WO_4. Induction periods (in seconds) for all three systems studied were well represented by the semiempirical equation

$$t_{ind} = kc_0^{-n} , \qquad (1)$$

where c_0 is the initial concentration (mol m^{-3}) of the precipitating substance. The constants k and n of eqn. (1), obtained by a least squares regression of the experimental data for a given substance and stirrer speed, are given in Table 1 together with the concentration ranges within which the correlation holds. For higher concentrations, precipitation is virtually instantaneous, whereas for lower ones the induction periods are long and indeterminate.

Inspection of Figure 1 reveals that, for Ag_2WO_4, the lower the initial concentrations and hence supersaturation, the more pronounced in absolute terms is the influence of mixing intensity on the induction period. The same pattern was followed by $SrWO_4$ and $CaWO_4$ (not shown here).

The shape of the dependence of induction period on stirrer speed for all three substances studied suggests that there is a limiting value of stirrer speed beyond which there is no further significant reduction in t_{ind} as shown in Figure 2 for $SrWO_4$. This limit was not actually reached in the present work, however, since at speeds above 700 rpm entrained air bubbles made turbidity observation unreliable and might also influence the nucleation process.

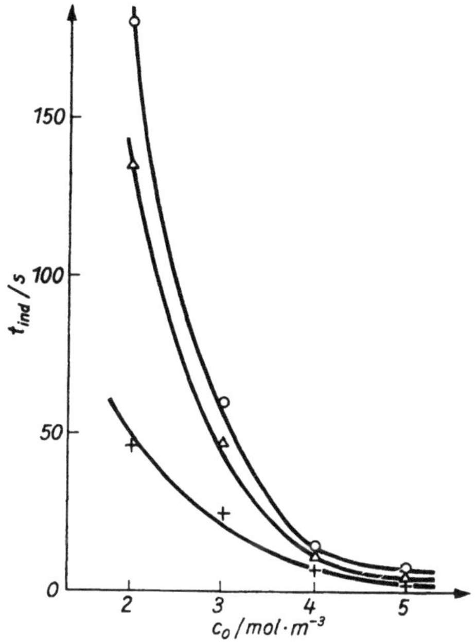

Fig. 1. Induction periods for Ag_2WO_4 at 25 °C as a function of initial concentration for various stirrer speeds

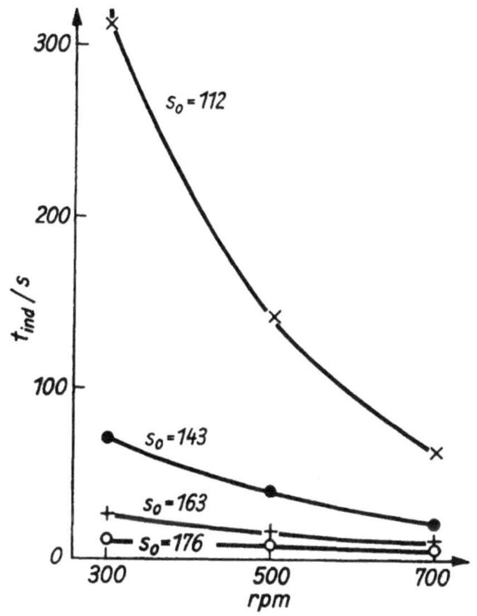

Fig. 2. Induction periods for $SrWO_4$ precipitation at 25 (°C as a function of stirrer speed for different levels of initial supersaturation S_0

Table 1
Experimental constants of eqn. (1) for different mixing intensities

Substance	rpm	n	k	N	r	c_0
Ag_2WO_4	300	3.65	$2.54 \cdot 10^3$	4	0.992	2—5
	500	3.30	$1.42 \cdot 10^3$	4	0.991	
	700	3.03	$4.77 \cdot 10^2$	4	0.974	
$CaWO_4$	300	1.28	$1.26 \cdot 10^3$	5	0.989	10—80
	500	1.27	$8.99 \cdot 10^2$	4	0.990	
	700	1.66	$2.25 \cdot 10^3$	6	0.995	
$SrWO_4$	300	4.96	$8.47 \cdot 10^8$	4	0.983	40—100
	500	4.17	$3.68 \cdot 10^7$	7	0.993	
	700	3.66	$3.54 \cdot 10^6$	6	0.940	

N — number of experimental points; r — correlation coefficient

Precipitation in this study was effected by mixing together equal volumes of aqueous solutions of Na_2WO_4 with $AgNO_3$, $CaCl_2$ and $SrCl_2$ respectively, producing final concentrations c_0 of tungstate and $2c_0$ of neutral electrolyte, $NaNO_3$ or NaCl. The equilibrium solubilities of Ag_2WO_4, $CaWO_4$ and $SrWO_4$ in water at 25 °C are reported as 0.5069, 0.1122 and 8.466×10^{-2} mol m^{-3} (LINKE). These values were corrected for the presence of the neutral electrolyte by employing activity coefficients calculated from the Güntelberg equation

$$\log \gamma_{\pm} = -0.509 \; |z_+ z_-| \; \sqrt{I}/(1 + \sqrt{I}) \, , \qquad (2)$$

where z_+ and z_- are the cation and anion charges, and the ionic strength of the solution $I = 0.5 \Sigma c_i z_i^2$. The initial level of supersaturation ($S_0 = c_0/c_{eq}$) was calculated from the initial concentration of the precipitating substance and the equilibrium solubility corrected as above. The ranges of supersaturation covered in the experimental work were 3.7—7.8, 32—134 and 121—210 for Ag_2WO_4, $CaWO_4$ and $SrWO_4$, respectively.

4. Discussion

An indication of the mechanisms controlling the induction period, i.e. a combination of nucleation and growth or growth alone, may be obtained by plotting the experimental data as

$$\log t_{ind} = A + BX \,, \tag{3}$$

where X represents $(\log S)^{-2}$ for homogeneous nucleation followed by diffusional growth, $(\log S)^{-1}$ for both mononuclear and polynuclear growth, or $\log \Delta c$ for screw-dislocation growth (SÖHNEL, MULLIN). Linearity of any of the plots suggests that the respective mechanism could be the controlling one.

Further pointers to the controlling mechanism may be provided by the constants A and B of eqn. (3) determined by data fitting. Since these constants have well-defined physical meanings, they can be predicted for each controlling mechanism and compared with the experimental values. Values of B with the exception of those for the screw-dislocation mechanism, may be used to calculate the interfacial tension γ^s (mJ · m^{-2}) which can then the compared with values estimated from (SÖHNEL 1983):

$$\gamma^s = -18.3 \log c_{eq} + 34.5 \,. \tag{4}$$

The expressions that can be used for such a procedure are as follows:

i) Homogeneous nucleation followed by diffusional growth (NIELSEN, SÖHNEL)

$$\gamma^s = 0.029(\nu^2 B/V_m^2)^{1/3} \,, \tag{5}$$

where V_m is the molar volume of the solid (m^3 mol^{-1}) and ν is the number of ions in a molecular unit.

$$A = -10 \,. \tag{6}$$

ii) Monomolecular growth (NIELSEN, SÖHNEL)

$$\gamma^s = 0.0339 \sqrt{\nu B}/V_m^{2/3} \,, \tag{7}$$

and

$$A = \log (V_m/4N_A D_s r_v) \,, \tag{8}$$

where D_s is the surface diffusion coefficient (m^2 s^{-1}), r_v is the size (m) of a detectable crystal, and N_A is the Avogadro number.

iii) Polynuclear growth (NIELSEN 1969)

$$\gamma^s = 0.0587 \sqrt{\nu B}/V_m^{2/3} \,, \tag{9}$$

and

$$A = \log r_v/(D_s V_m^{1/3} \Delta c^{2/3} N_A^{1/3}) \,, \tag{10}$$

where $\Delta c = (c_0 - c_{eq})$ (mol m^{-3}).

iv) Screw-dislocation growth (SÖHNEL, MULLIN)

$$B = -g \,, \tag{11}$$

and

$$A = r_v/k_g \,, \tag{12}$$

where g is the kinetic order of growth and k_g is the growth constant.

Based on the linearity of the different plots of eqn. (3) for the three tungstates studied, within the investigated range of supersaturation, screw-dislocation growth can be identified as the most likely controlling mechanism of the induction period for Ag_2WO_4, whereas this mode is rejected for $SrWO_4$ (see Tab. 2). All the other plots for $SrWO_4$ and $CaWO_4$ give equally good linearity, so the controlling mechanism cannot be inferred from them, and it is necessary to compare experimental and predicted values of the constants of eqn. (3)

Experimentally determined and expected (quoted in brackets) values of A and γ^s are given in Table 2. For calculation purposes, the following numerical values were used: $\varrho(SrWO_4) = 6184\ kg\ m^{-3}$, $\varrho(CaWO_4) = 6062\ kg\ m^{-3}$ (GMELIN), $\nu = 2$, $r_v = 10^{-6}$ m, $D_s = 10^{-9}$ to $10^{-12}\ m^2\ s^{-1}$ (NIELSEN 1969).

Table 2
Comparison of experimental and predicted values of A, γ^s and g

X in eq. (3)	$(\log S)^{-2}$ nucleation/growth	$(\log S)^{-1}$ mononuclear	$(\log S)^{-1}$ polynuclear	$\log \Delta c$ screw-dislocation
plot	L	L	L	C
$SrWO_4$ —A	6.2 (10)	13.5 (12.1)	13.5 (3.0)	—
γ^s	106 (107)	60 (107)	104 (107)	—
plot	L	L	L	L
$CaWO_4$ —A	0.6 (10)	2.7 (12.2)	2.7 (2.8)	1.1 (?)
γ^s	68 (109)	102 (109)	176 (109)	—
g	—	—	—	1.4 (1)
plot	C	C	C	L
Ag_2WO_4 —A	—	—	—	5.47 (?)
g	—	—	—	2.6 (2)

The expected A values (in parentheses) are the arithmetic means of calculations using values of $D_s = 10^{-9}$ and 10^{-12}. Plots are described as either linear (L) or curved (C).

Inspection of Table 2 reveals that the most likely controlling mechanism for $SrWO_4$ is homogeneous nucleation followed by diffusion controlled growth, since only in this case (see column 2) do both the A and γ^s experimental values compare well with the expected values. Screw-dislocation growth can be identified as the controlling mechanism for $CaWO_4$ since all other mechanisms that also give linear plot are excluded by the disagreement between the experimental and expected values of A and γ^s. In the case of screw-dislocation growth the constant A in eqn. (3) cannot be predicted for either $CaWO_4$ or Ag_2WO_4 as the respective k_g values are unknown. Experimental values of g, however, compare favourably with those expected (1 and 2) for growth at high and low supersaturation respectively (NIELSEN 1984) as is indeed the case of $CaWO_4$ and Ag_2WO_4 in this study.

Thus, based on the above analysis, the three tungstates studied can be divided into two categories according to the mechanism controlling the induction period:

(i) $SrWO_4$: homogeneous nucleation followed by diffusional growth,

(ii) $CaWO_4$ and Ag_2WO_4: screw-dislocation growth of crystals formed at the very beginning of the precipitation.

None of the processes of homogeneous nucleation, diffusion-controlled growth of very small crystals or screw-dislocation growth, can be significantly influenced by mixing. But mixing was found to exert a considerable influence on the precipitation of all three substances studied. One possible conclusion, therefore, it that either a supersaturation build-up at the very beginning of the precipitation process (immediately after mixing the reacting solutions) or crystal agglomeration, or both, must be accelerated by mixing and thus be responsible for the induction period decrease when the intensity of mixing increases.

Mixing exerts a profound influence on the batch precipitation of sparingly soluble substances which becomes more significant in absolute terms as the initial supersaturation is lowered. The behaviour of sparingly soluble substances is thus comparable with that already reported for the crystallization of soluble substances (Garside et al.; Mullin, Osman; Mullin, Žáček).

One of us (O.S.) is indebted to the Science and Engineering Research Council of Great Britain for financial support.

References

Chianese, A., di Cave, A., Mazzarotta, B.: in: Industrial Crystallization 84, eds. S. J. Jančić, E. J. de Jong, Amsterdam 1984, p. 443
Garside, J., Gaska, C., Mullin, J. W.: J. Cryst. Growth **13/14**, 510 (1972)
Gmelin: Handbuch der Anorganischen Chemie, 8. Auflage, No 54, Berlin 1933, pp. 290, 296
Jančić, S. J., Grootscholten, P. A. M.: Industrial Crystallization, Delft 1984
Linke, W. F.: Solubilities of Inorganic and metal-organic Compounds, New Jersey 1958
Liu, S-T., Nancollas, G. H., Gaseicki, E. A.: J. Cryst. Growth **33**, 11 (1976)
Mullin, J. W.: Crystallization, 2nd ed. London 1972
Mullin, J. W., Osman, M. M.: Kristall u. Technik **8**, 471 (1973)
Mullin, J. W., Žáček, S.: J. Cryst. Growth **53**, 515 (1981)
Nývlt, J., Skřivánek, J., Gottfried, J., Křičková, J.: Coll. Czech. Chem. Commun. **31**, 2127 (1966)
Nývlt, J., Söhnel, O., Matuchová, M., Broul, M.: Kinetics of Industrial Crystallization, Amsterdam 1985
Nielsen, A. E., Kristall u. Technik **4**, 17 (1969)
Nielsen, A. E., Söhnel, O.: J. Cryst. Growth **11**, 233 (1971)
Nielsen, A. E.: in Industrial Crystallization 81, ed. S. J. Jančić, E. J. de Jong, p. 35, Amsterdam 1982
Nielsen, A. E., J. Cryst. Growth **67**, 289 (1984)
O'Hern, H. A., Rush, F. E.: IEC Fundamentals **2**, 267 (1963)
Pohorecki, R., Baldyga, J.: Chem. Eng. Sci. **38**, 79 (1983)
Pohorecki, R., Baldyga, J.: 6th Europ. Conf. on Mixing, Paper 12, Würzberg, F. R. Germany 1985
Roquez, H., Girou, A.: Water Res. **8**, 907 (1974)
Söhnel, O.: Cryst. Res. Technol. **16**, 651 (1981)
Söhnel, O.: J. Cryst. Growth **63**, 174 (1983)
Söhnel, O., Mullin, J. W.: to be published
Takubo, H.: J. Cryst. Growth **72**, 631 (1985)
Turnbull, P.: Acta Metallurg. **1**, 684 (1953)

(Received March 25, 1987)

Authors' addresses:

O. Söhnel
Research Institute of Inorganic Chemistry
Revoluční 86
400 60 Ústí nad Labem, Czechoslovakia

Prof. J. W. Mullin
University College London
Department of Chemical and Biochemical Engineering
Torrington Place
London WC1E 7JE, England

Cryst. Res. Technol.	22	1987	10	1241–1248

Christo BALAREW

Inorganic Salts Research Laboratory
Institute of General and Inorganic Chemistry, Bulgarian Academy of Sciences, Sofia

Evaporite Minerals Succession

An explanation is given for the crystallization succession during sea-water evaporation, namely for the differences between the equilibrium solubility diagram of the marine-type system and the "solar diagram": The magnesium ions are known to exist in sea-type solutions predominantly as hexaaqua complexes. Highest crystallization rates and lowest supersaturations required for nucleation should be exhibited by the magnesium salts, crystal structures of which incorporate unchanged hexaaqua magnesium complexes. For the salts with crystal structures in which the water molecules from the magnesium hexaaqua ions are partially or completely replaced by other ligands, the nucleations require high supersaturations and for this reason they do not appear in the "solar diagram".

This model is used to explain also some genetic pecularities of evaporite deposits: primary evaporite minerals are those which occur according to the "solar diagram" of crystallization. The evaporite minerals featuring in the equilibrium diagram and absent in the "solar diagram" occur in nature most frequently as secondary minerals.

Introduction

The onset of the contemporary stage in the development of the notions about the formation of evaporite minerals is connected with the classical works of VAN'T HOFF and coworkers in the years 1897–1908.

VAN'T HOFF (1905; 1909) has assumed that the formation of natural salt deposits is related to sea water evaporation. For this reason he considered that the study on the sequence of salt crystallization during sea water evaporation may provide accurate information of evaporite minerals location, the paragenetic ratios and hence may help elucidate the conditions and relationships governing their formation. To that purpose VAN'T HOFF had studied the stable equilibria in the fivefold sea water-type system Na^+, K^+, Mg^{2+}/Cl^-, SO_4^{2-}/H_2O at 25 to 83 °C temperatures. By introducing calcium salts in these systems he studied also conditions for the formation of a number of calcium minerals. The equilibrium diagram obtained by VAN'T HOFF for the five-fold system in its cross-section for saturated sodium chloride solutions is displayed in Figure 1. The salt crystallization succession for the particular case of sea water at ordinary temperature is exhibited schematically in Figure 2a.

A number of peculiar features of the natural deposits, however, turned out to deviate substantially from the VAN'T HOFF scheme. Thus, for example, in the Zechstein salt deposits in Germany, primary kainite is absent, there exists the paragenesis sylvite-kiserite, no bloedite, epsomite and hexahydrite are found.

To explain this contradiction ARRHENIUS and LACHMANN created the theory of thermometamorphism. According to this theory, the primary evaporite minerals, encased by subsequent deposits, sink to a depth down to 2–3 km where their temperature is raised to 80 °C and higher. Under these conditions a number of crystal

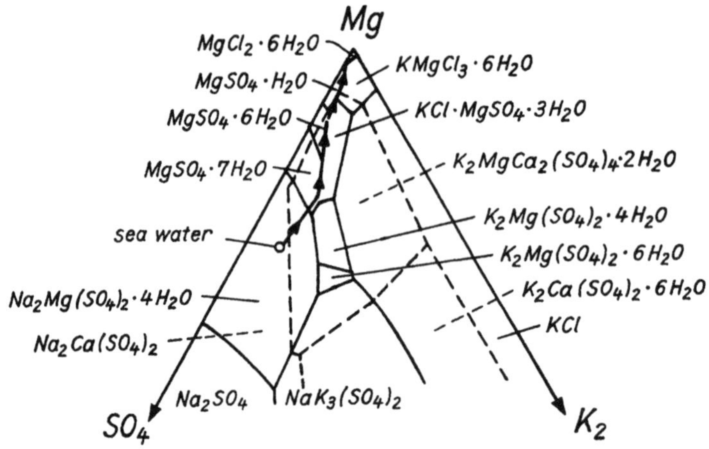

Fig. 1. Equilibrium diagram of the five-fold Na^+, K^+, Mg^{2+}/Cl^-, SO_4^{2-}/H_2O system at 25 °C in its cross-section for sodium chloride saturated solutions. Dotted lines are for Ca^{2+}-containing evaporites

hydrates such as gypsum, epsomite, kainite, and others, dehydrate and transform into anhydrite, kieserite, etc, respectively. The released crystal water produces brines which on their turn alter additionally the mineral composition. This theory was elaborated further by other researchers and, with some corrections, is accepted by many scientists.

In 1927 KURNAKOW and NIKOLAEW began studies on the evaporation of the saline Sax lake which intakes water from the Black Sea. These studies brought about the discovery of the so called "solar succession" for salt crystallization from sea water (Fig. 2b) as well as to the "solar diagram" of evaporation (Fig. 3). The "solar diagram" reflects equilibria in the metastable region. The metastable salts, however, do exist because the sea-water-type solutions can remain persistently supersaturated with respect to some of the stable phases. Consequently, the "solar diagram" reflects the kinetics of crystallization and phase transitions. As seen from Figures 2b and 3, there are no crystallization fields for the bloedite and kainite as well for kieserite in the "solar diagram". A crystallization field of sylvite, as a primary evaporite mineral, is seen to appear. The finding that during sea water evaporation, sylvite instead of kainite is obtained, was reported also by VAN'T HOFF (1899) in his earlier works.

Although the "solar succession" of crystallization may be used to explain considerably better some peculiarities of natural marine salt deposits, several problems remain obscure:

According to the "solar succession", during the first stages of sea water evaporation, after separation of calcium and magnesium carbonates, the precipitation of gypsum follows. All the evaporation attempts with different sea waters at temperatures 20 to 25 °C carried out thus far yield gypsum and no anhydrite. In the mineral evaporite deposit of marine origin, however, anhydrite is predominantly found whereas gypsum is rarely met as primary deposit, and it is not a rock-forming mineral (IVANOW). Probably the past day crystallizations from marine salt-generating basins have produced most frequently gypsum; the anhydrites, being gypsum dehydration products are secondary formations (STRAKHOW).

In the majority of evaporite formations of marine origin, magnesium sulphate is present only as kieserite, whereas namely kieserite should not occur as a mineral

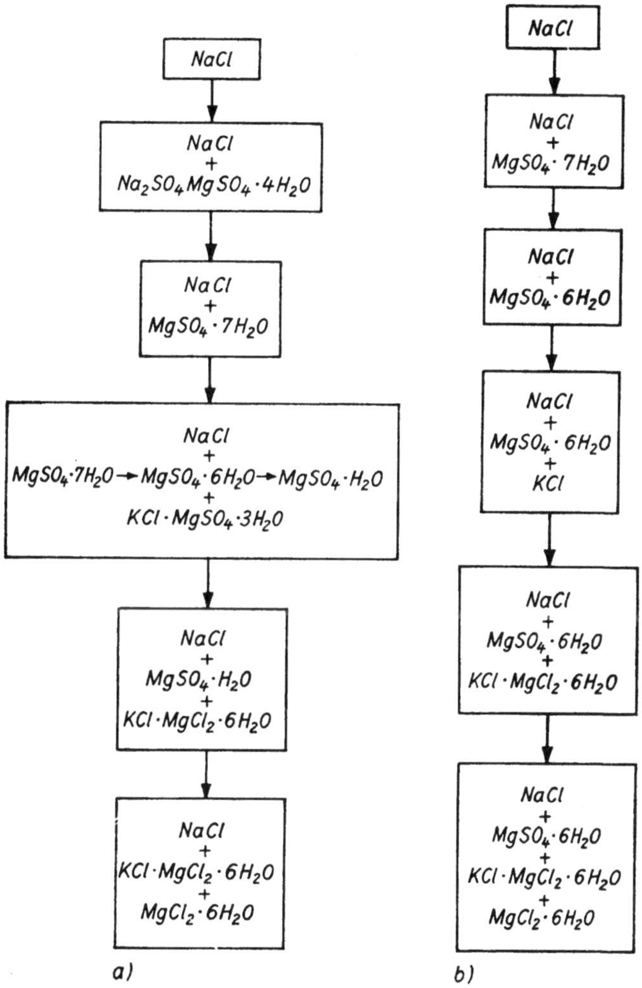

Fig. 2. Salt crystallization succession by sea-water evaporation: (a) according to the equilibrium diagram; (b) according to the "solar diagram"

generated according to the "solar diagram". The "solar succession" of crystallization cannot explain also the presence of kainite deposits, many of which are even of industrial importance. Anyway, there is ample evidence that these two minerals, similarly to anhydrite, are most frequently of secondary origin (IVANOW).

Purposes of this paper:

i. to consider in detail the reasons for the differences in the salt crystallisation successions in the cases of the equilibrium and of the "solar diagrams", and

ii. to explain some differences between the peculiarities of natural evaporite deposits of marine origin and the parageneses defined by the "solar succession" for salt crystallization.

The present work is based on the five-fold system Na^+, K^+, Mg^{2+}/Cl^-, SO_4^{2-}/H_2O hence Ca^{2+}-containing evaporites are outside of consideration. (Data on such evaporites have been published by H. P. EUGSTER, C. E. HARVIE & J. H. WEARE: Geochim. & Cosmochim. Acta **44**, 1335–1348 (1980).

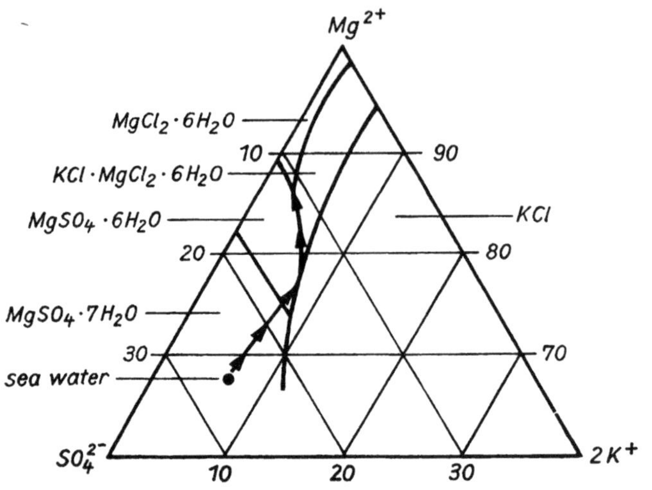

Fig. 3. "Solar diagram" for the five-fold Na^+, K^+, Mg^{2+}/Cl^-, SO_4^{2-}/H_2O system at 25 °C in its cross-section for sodium chloride saturated solutions (the region of magnesium chloride-rich solutions)

A model for the kinetics of crystallization of the magnesium salts

The magnesium ions are known to exist in sea-type waters at ordinary temperatures in very wide concentrational ranges (from dilute to comparatively highly concentrated solutions) predominantly as hexaaqua complexes, $Mg(H_2O)_6^{2+}$. According to the concept of hard and soft Lewis acid and bases, they are the most stable complexes as compared with all other complexes that may exist in these solutions (Balarew et al. 1984). The highest crystallization rate, and the lowest supersaturation required for nucleation, should be exhibited by the salts, whose crystal structures incorporate unchanged hexaaqua magnesium complexes, and these are namely $MgSO_4 \cdot 7\,H_2O$ (Baur), $MgSO_4 \cdot 6\,H_2O$ (Zalkin et al.), $KCl \cdot MgCl_2 \cdot 6\,H_2O$ (Fischer) and $MgCl_2 \cdot 6\,H_2O$

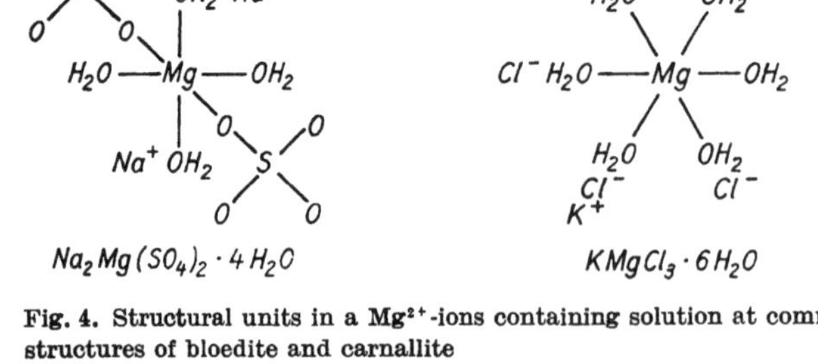

Fig. 4. Structural units in a Mg^{2+}-ions containing solution at common temperature and in the structures of bloedite and carnallite

(ASLANIAN et al.). For salts in the crystal structures of which the magnesium ions are coordinadet not only by water molecules but also by other ligands (Fig. 4), their nucleations require higher supersaturations and the spontaneous crystallization processes do not proceed. Further water evaporation results in more concentrated solutions thus increasing the ionic strength of the solutions and decreasing the water activity, which at sufficiently high concentrations governs the appearance in the solution, parallel with the magnesium hexaaqua ions, of appreciable quantities of other water-deficient octahedral magnesium complexes. Thus, the supersaturation required for the crystallization of these magnesium salts (its crystal structure contains magnesium complexes with less water molecules) will decrease with increasing concentration of the solutions, for example, in the series: $Na_2SO_4 \cdot MgSO_4 \cdot 4\,H_2O$, $KCl \cdot MgSO_4 \cdot 3\,H_2O$, $MgSO_4 \cdot H_2O$, $K_2SO_4 \cdot 2\,MgSO_4$.

In support of the above claim, a number of data may be cited. Thus, for example, EMONS and STEGMANN when studying chosen solutions of the marine-type five-fold system at 85 °C have found that the trend towards langbeinite and kainite crystallization appears above a given, comparatively very high magnesium sulphate and magnesium chloride concentrations in the solution and increases with increasing concentration. Moreover, these researches have found that intermediate compositions of the solutions exist at which only kainite and no langbeinite are found to crystallize.

These results as well as many other show that for the crystallization of the lower hydrates of the simple and double magnesium salts, a very high supersaturation is necessary; such high supersaturations cannot be obtained in solar evaporation since much lower supersaturations suffice to cause the crystallization of salts whose crystal structu res contain the unit $Mg(H_2O)_6^{2+}$. These latter salts may not be the thermodynamically stable solid phases under the corresponding conditions and compositions of the solutions. Their formation, however, is due to the fact that their solubility under these conditions is not much higher than that of the stable salts.

In this way, by using the model given here for the relationship between the structure of the water solution and the structure of the crystal hydrates as well as the interpretation on its basis of the kinetics of crystallization of the different simple and double magnesium salts, an explanation is found for the differences between the equilibrium solubility diagram of the five-fold marine-type system and the "solar diagram".

Genetic peculiarities of evaporite mineral deposits

The model stated above may be used to explain also some genetic peculiarities of salt deposits. First, this is the appearance of sylvite as a primary evaporite mineral which is related to the much higher supersaturations required for the kainite crystallization as compared with that of sylvite from the corresponding solutions at an ordinary temperature. It is also known that bloedite is met rarely in mineral deposits and it does not from considerable deposits (IVANOW). The same refers also to langbeinite the deposits of which are of a predominant mineralogical importance (IVANOW).

The problem of kieserite and kainite deposits, however, is quite different. These two salts, due to the high supersaturation needed for their crystallization are not present in the "solar diagram" of crystallization. They form, however, deposits of industrial importance. As noted already, these two minerals are considered to have most frequently a secondary origin, resulting from syngenetic metamorphism (IVANOW). Their genesis can be explained using VALJASHKO's theory for the formation of potassium deposits.

VALJASHKO starts with the finding that the potassium salt crystallization from sea-type brines begins when the volume of the salts having crystallized become equal or even larger than the volume of the remaining mother solution. Then, a considerable part of the brine remains distributed among the pores between the salt crystals already formed. The separation of this brine from the salt bulk may occur, if, as a result of sinking of some parts of the evaporite basin, conditions are created for its drawing out. The brine collected in a basin of open halite deposits continues to be evaporated further, making possible the potassium salts crystallization.

It is seen from the "solar diagram" that the potassium salt crystallisation begins when the main bulk of epsomite and hexahydrite has already been separated. The drawing out of the highly concentrated potassium brines would occur mainly through them. The water vapour pressure above these brines is below 9 mm Hg at 20 °C (TRENDAFELOV et al.). By using the solubility diagrams for the systems $KCl–MgSO_4–H_2O$ (KURNAKOW and SCHOYCHET 1938) and $MgSO_4–H_2SO_4–H_2O$ (BALAREW, MARKOV) it may be calculated from the eutonic points with the $MgSO_4 \cdot 6\,H_2O$ and $MgSO_4 \cdot H_2O$ participation that the hexahydrite dehydration to kieserite requires at 25 °C a water vapour pressure above the saturated solution in equilibrium with these crystal phases of about 10 mm Hg. Consequently, the solution in contact with the hexahydrite and epsomite crystals would be able to dehydrate these crystals yielding kieserite.

As far as the genesis of kainite deposits, KURNAKOW and NIKOLAEW's studies have shown that kainite is formed in the process of carnallite's syngenetic recrystallization in the presence of solutions containing magnesium sulphate:

$$KCl \cdot MgCl_2 \cdot 6\,H_2O + \underset{\text{(solution)}}{MgSO_4} = KCl \cdot MgSO_4 \cdot 3\,H_2O + \underset{\text{(solution)}}{MgCl_2} + 3\,H_2O$$

An overall picture for the genesis of evaporite minerals by solar sea water evaporation is schematically presented in Table 1.

Table 1
Evaporite minerals succession

Zone	primary rock-forming minerals	secondary and accompanying minerals (diagenetic and epigenetic)
1. gypsum-anhydrite	$CaSO_4 \cdot 2\,H_2O$	$CaSO_4$
2. halite	NaCl	$CaSO_4$
	$CaSO_4 \cdot 2\,H_2O$	$K_2SO_4 \cdot MgSO_4 \cdot 2\,CaSO_4 \cdot 2\,H_2O$
3. magnesium sulphate	$MgSO_4 \cdot 6\,H_2O$	$MgSO_4 \cdot H_2O$
	$MgSO_4 \cdot 7\,H_2O$	$K_2SO_4 \cdot MgSO_4 \cdot 2\,CaSO_4 \cdot 2\,H_2O$
	NaCl	$Na_2SO_4 \cdot MgSO_4 \cdot 4\,H_2O$
4. sylvite	KCl	$KCl \cdot MgSO_4 \cdot 3\,H_2O$
	$MgSO_4 \cdot 6\,H_2O$	$K_2SO_4 \cdot 2\,MgSO_4$
	$MgSO_4 \cdot 7\,H_2O$	$MgSO_4 \cdot H_2O$
	NaCl	$K_2SO_4 \cdot MgSO_4 \cdot 4\,H_2O$
5. carnallite	$KCl \cdot MgCl_2 \cdot 6\,H_2O$	$MgSO_4 \cdot H_2O$
	$MgSO_4 \cdot 6\,H_2O$	$K_2SO_4 \cdot 2\,MgSO_4$
	NaCl	$KCl \cdot MgSO_4 \cdot 3\,H_2O$
6. bischofite	$MgCl_2 \cdot 6\,H_2O$	$MgSO_4 \cdot H_2O$
	$KCl \cdot MgCl_2 \cdot 6\,H_2O$	
	$MgSO_4 \cdot 6\,H_2O$	
	NaCl	

The treatment reported here and based on the studies on evaporite mineral formation from sea water shold in principle be also valid in the cases of differently metamorphosed continental waters. For these waters differences in the crystallization successions defined by the equilibrium diagram and by the corresponding "solar diagram" of crystallisation is also expected to be present. Starting with the equilibrium diagram and using the consideration presented in this work for the relation between the structure of the solution and the supersaturation needed to obtain a given crystal phase, it is possible to predict the corresponding "solar succession" of crystallization and hence to explain the evaporite mineral formation from correspondingly metamorphosed continental waters.

Conclusions

The approach elaborated here for explaining the differences in the salt crystallization successions at equilibrium and by solar evaporation on the one side, and of the real natural parageneses on the other hand, can be used to draw the following general conclusion: In nature, primary evaporite minerals are those which occur according to the "solar diagram" of crystallisation. The minerals which are thermodynamically stable under the corresponding conditions and in the respective medium and absent in the "solar diagram" of crystallisation occur in nature most frequently as secondary minerals, resulting from syngenetic, diagenetic or epigenetic processes.

The author is grateful to Dr. Rafael Rodriguez Clemente, Madrid, who promptly recommended to write this paper, and to Professor Stoian Trashliev, Sofia, for his help and useful discussions.

References

ARRHENIUS, S., LACHMANN, R.: Geol. Rundschau **3**, 139 (1912)
ASLANIAN, S., BALAREW, Chr., SPASSOV, D.: Kristall und Technik **13**, 1451 (1978)
BALAREW, Chr., DUHLEV, R.: J. Solid State Chem. **55**, 1 (1984)
BALAREW, Chr., DUHLEV, R., SPASSOV, D.: Crystal Res. Technol. **19**, 1469 (1984)
BALAREW, Chr., MARKOV, L.: Jahrb. Chem.-Technol. Hochschule, Burgas **10**, 607 (1973)
BAUR, W. H.: Acta Cryst. **17**, 1361 (1964)
EMONS, H.-H., STEGMANN, E.: Freib. Forschungs-H. **A436**, 65 (1968)
FISCHER, W.: N. Jb. Miner. Mh. **3**, 100 (1973)
IVANOW, A. A.: Fundamentals of Geology and Methods of Searching, Investigating and Evaluating Mineral Salt Deposits, Moscow 1953, p. 10 and 110
KURNAKOW, N. S., SCHOYCHET, D. N.: Equilibrium in the Reciprocal System Potassium Chloride — Magnesium Sulphate at 25 °C. Reports of the Division for Physicochemical Analysis, Academy of Sciences USSR **10**, 305 (1938)
KURNAKOW, N. S., NIKOLAEW, V. I.: Solar Evaporation of Sea Water and Lake Brines. Reports of the Division for Physicochemical Analysis, Academy of Sciences USSR **10**, 333 (1938)
STRAKHOW, N. M.: Fundamentals of the Theory of Lithogenesis. Vol. III, Acad. Sci. USSR, Moscow (1962), p. 443
TRENDAFELOV, D., OBRETENOV, Z., BALAREW, Chr., KRISTANOVA, L., MILOSCHOVA, M., OIKOVA, T., KARAIVANOVA, V.: Chemie und Industrie, Sofia **43**, 99 (1971)
VALJASHKO, M. G.: Geochemical Relationships of the Formation of Potassium Salt Genesis, Moscow 1962, p. 194

VAN'T HOFF, J. H.: Zur Bildung ozeanischer Salzablagerungen, Bd. I und II, Braunschweig 1905, 1909
VAN'T HOFF, J. H., MEYERHOFFER, W.: Untersuchungen über die Bildungsverhältnisse der oceanischen Salzablagerungen, insbesondere des Stassfurter Salzlagers. XIII. Das Eintrocknen des Meerwassers bei 25°. S.-Ber. Akad. Wiss., Berlin 1899, p. 372
ZALKIN, A., RUBEN, H., TEMPLETON, D. H.: Acta Cryst. **17**, 235 (1964)

(Received April 4, 1987)

Author's address:

Dr. Christo BALAREW
Institute of General and Inorganic Chemistry
Bulgarian Academy of Sciences
1040 Sofia
Bulgaria

Cryst. Res. Technol.	22	1987	10	1249–1254

M. BETZL, E. N. VLASOVA, N. B. DJAKONOVA, V. I. MATORIN

Akademie der Wissenschaften der DDR, Zentralinstitut für Kernforschung, Rossendorf, and Central Research Institute for Ferrous Metallurgy, Moscow, USSR

Effect of Structural Transformations on the Texture of Ni–Fe Base Alloys

1. Processes of texture formation were studied in three Ni–Fe system alloys containing Mo and Ti (alloy No. 1), Cr, Ti and Al (alloy No. 2) and Co, Ti and Al (alloy No. 3). The deformation texture is identical in all alloys which agrees with the proximity of stacking fault energies as estimated from X-ray data.
2. Annealing leads to a pronounced diffusion and alteration of the texture in alloy No. 1, but does not alter the texture type and intensity in alloy No. 2, whereas in alloy No. 3 the typical strain texture components are intensified.
3. Structure processes governing texture formation were identified as recrystallization in alloy No. 1, and oriented $\gamma - \gamma'$ transformation in alloys No. 2 and 3 where the intergranular boundaries were blocked by heterogenous precipitation of phases not possessing fcc structure.

Introduction

Nickel-iron base alloys represent a family of commercial metals possessing specific magnetic and elastic properties. The basic elements used in industry for alloying this binary system are Co, Ti, Al, Cr, and Mo, the result being various phase transformations which enable the principal magnetic and elastic properties to be controlled, along with improving the strength characteristics. The actual levels of physical properties of commercial specimens depend to a substantial degree, owing to anisotropy, on alloy texture, i.e. on the spacial distribution of the orientation of crystallites constituting the specimen. Studies dealing with the texture formation in Ni–Fe base alloys affected by deformation and phase transformations are far from exhaustive, and particularly lacking is the application of modern techniques for the analysis of orientation distribution functions (ODF).

The present study was aimed at investigating the effect of alloying on the formation of deformation texture in Ni–Fe base fcc alloys, and at gaining an insight into the texture changes during annealing processes.

Materials and techniques

Three commercial alloys were studied (see table): a magnetically soft alloy (No. 1) rich in nickel and alloyed with Ti and Mo; elinvar (No. 2) alloyed with Cr, Ti and Al, having an almost zero elastic modulus at room temperature; and a high-temperature elinvar (No. 3) alloyed with Co, Ti and Al (VLASOVA 1980, 1984). All specimens were subjected to 75 and 90-percent cold deformation followed by annealing for 20 hrs at 700 °C.

The main instrumental techniques used were X-ray and neutron diffraction analyses. Alloy texture was studied by the Schultze X-ray diffraction method utilizing Co radiation.

Alloy	alloy composition at. percent						
No	Ni	Fe	Co	Cr	Ti	Al	Mo
1	69	23	—	—	3	—	5
2	43	47	—	5.5	3.5	1	—
3	29	41	25	—	3	2	—

Partial pole figures were obtained within $0 < \alpha < 80°$, with 5° intervals for both α and β angles. The orientation distribution functions were calculated from 111, 100, 110, and 311 pole figures (Bunge). Post-annealing structures were studied using X-ray techniques in vacuum with photorecording of monochromatic CoK_α radiation.

Experimental results

The cold deformation texture of all alloys is characterized by a continuous distribution of orientations. The probability is higher along skeleton lines which are roughly similar in position for the three alloys under study. In Figure 1a are shown the skeleton lines in Eulerian angle space (φ_1, Φ, φ_2) describing the rotation of crystal axes system relative to the specimen coordinates system (rolling direction and the two axes square to it). Within the "a—b" section the change of orientation involves the change of rolling direction from ⟨001⟩ to ⟨112⟩, the rolling plane (110) being constant. The

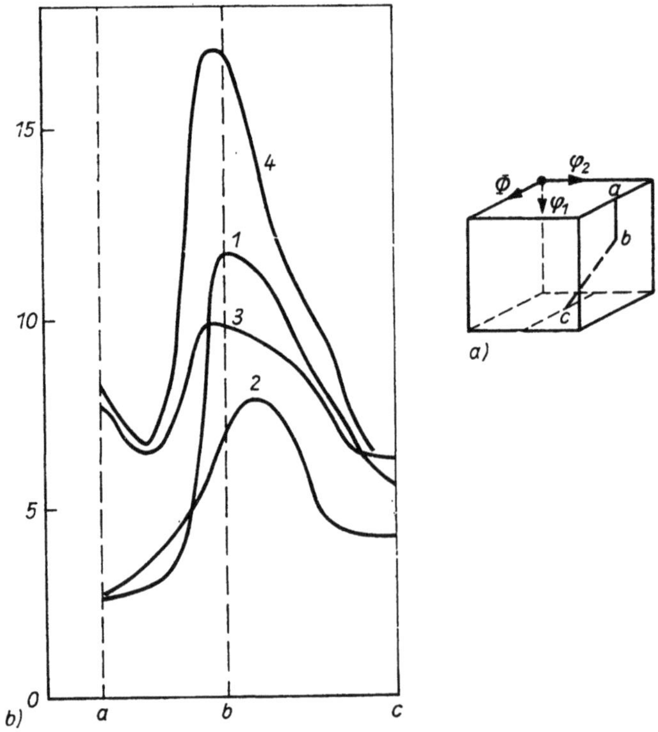

Fig. 1. Skeleton lines (a) in Eulerian angle space for fcc alloys and orientation distributions (b) along "a—b—c" skeleton lines for cold-strained alloys: 1 — Alloy No. 1, 75 percent strain; 2 — Alloy No. 2, 75 percent strain; 3 — Alloy No. 3, 75 percent strain; 4 — Alloy No. 3, 90 percent strain

change of the orientation distribution function $f(\varphi_1, \Phi, \varphi_2)$ along the "a—b—c" skeleton line is shown in Figure 1b for the 75-percent cold deformed alloys.

The distribution function for the cobalt containing alloy (No. 3) is distinguished by a weak maximum at point "a" and a larger flattened peak at point "b". The increase of deformation markedly intensifies the second peak and reduces its half-width. Predominant becomes the texture component with (110) plane aligned with the rolling plane and the ⟨112⟩ axes aligned with the rolling direction (see Fig. 1b).

The alloy containing Ti and Mo (No. 1) is characterized by a single maximum at point "b" of the same skeleton line. This peak is skewed, the probability density gradually decreasing toward point "c".

The single peak typical of the alloy with Cr, Ti, and Al (No. 2) is also located close to point "b" and somewhat skewed toward point "c".

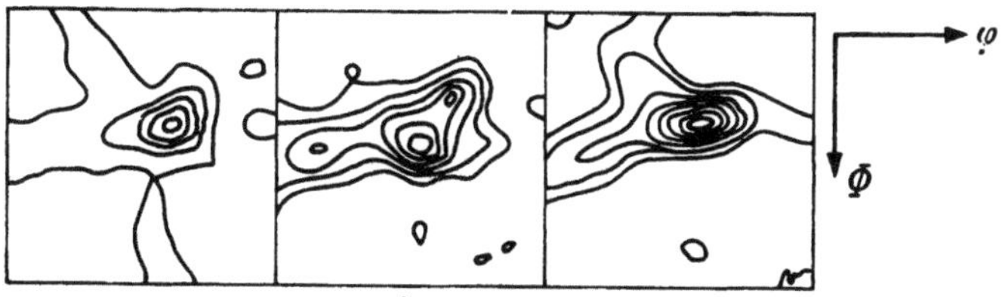

Fig. 2. Sections of orientation distribution functions at $\varphi_2 = 70°$ for alloy No. 1 (a), alloy No. 2 (b) and alloy No. 3 (c)

Although the distribution of crystallite orientation in the three alloys studied is basically identical, there are some features peculiar to each of the alloys, as can be seen from Figure 2 which presents the ODF sections at $\varphi_2 = 70°$. The ODF for the chromium containing alloy (No. 2) differs substantially in that the skeleton line is split. Estimates were made to compare the stacking fault energies, considering such as one of the principal factors affecting the deformation mechanism and, consequently, the type of texture. The energies were estimated from X-ray data basing on the shift of 111, 200, and 220 lines and by comparing the integral half-widths of 111, 200, 222, and 400 lines. The results displayed no differences, within error limits, between the stacking fault energies, which is in agreement with the overall similarity of the textures of the three alloys. It is possible that increasing the accuracy of energy measurements might enable differences to be established and matched with the respective variations in the details of orientation distribution functions.

Discussion

Let us consider the structural transformations during annealing and their effect on texture changes. The processes involved vary with the alloying elements and range from recrystallization to solid solution decomposition yielding different types of precipitates.

Alloy No. 1. The dotted diffraction lines on the X-ray patterns of the alloy with Ti and Mo after annealing for 20 hrs at 700 °C (see Fig. 3a) suggest that recrystallization is the basic process during annealing of this alloy. The lines attributed to fcc solid solution are accompanied by a series of very weak lines typical of hcp (η) phase based on Ni_3Ti, though present in a minor amount (ca. 2 percent). In Figure 4a are shown the changes of ODF along the "a—b" skeleton line and perpendicular to it at

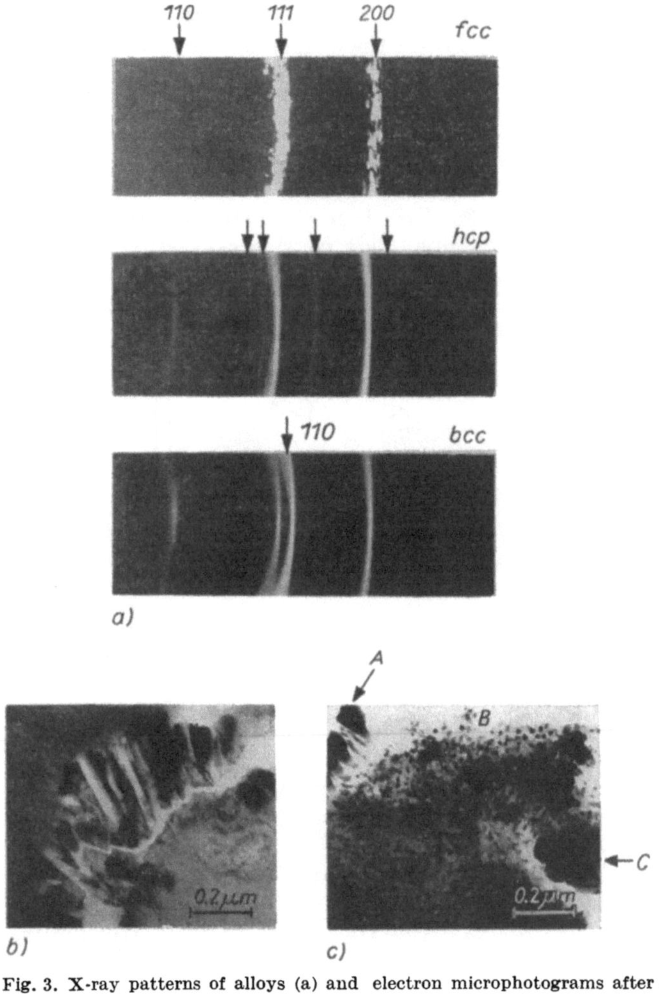

Fig. 3. X-ray patterns of alloys (a) and electron microphotograms after annealing for 20 hrs at 700 °C for alloy No. 2 (b) and alloy No. 3 (c)

point "b" in the cold-strained and annealed states. Apparently, recrystallization leads to considerable diffusion of the texture, and the new grain nuclei are not coordinated with the orientation of the initial crystallites. It seems that under the given conditions of annealing, nucleation occurs at the most distorted locations.

Alloy No. 2. As seen in Figure 3a, annealing of the alloy with Cr and Ti results in narrowing of the diffraction lines, indicating the elimination of the majority of strain induced faults. However, intensive grain growth is absent as would be otherwise evident from a dotted pattern of diffraction lines. In addition to fcc solid solution lines, the X-ray pattern displays two series of diffraction lines which are an evidence of two reactions taking place during annealing: the formation of a hexagonal η-phase based on Ni_3Ti similar to that in the alloy with Ti and Mo, and the appearance of a type LI_2 ordered γ'-phase. In this case the volume fraction of the hcp phase is higher (ca. 5 to 8 percent) compared to alloy No. 1. It has been shown (VLASOWA 1980) that the precipitation of the ordered γ'-phase, as the prevailing process, may follow two mechanisms: homogenous and cellular decomposition, the second mechanism dominating in the case of annealing after cold strain. The homogenous transformation involves the formation of equiaxial coherent γ'-phase particles within the grains, particle orientation being identical to that of the initial matrix grains: the type of texture is maintained.

In the case of cellular decomposition the orientation of the lamellae in colonies growing out of definite zones on grain boundaries (Fig. 3b) coincides with the orientation of one of the matrix grains. Obviously, this process might significantly affect the alloy texture depending on process conditions. The effect of this reaction on the alloy texture is illustrated in Figure 4b. The orientation distribution function maintains its basic features typical of the cold-deformation alloys. The maximum at point "b" along the "a—b" line becomes more pronounced and its half-width is reduced. It can be thus concluded that recrystallization and phase transformations do not lead to diffusion and changes of texture.

Alloy No. 3. The alloy with Co and Ti also undergoes two reactions during annealing. One of these involves, as in the previous case, the formation of a LI_2 ordered γ' phase based on Ni_3Ti. The second reaction is the precipitation and subsequent ordering of a bcc phase with a composition close to stoichiometric CoFe. The γ—γ' transformation is principally identical to that of the alloy with Cr and Ti (No. 2). Similarly, two transformation mechanisms are observed, the heterogenous reaction determining the formation of alloy texture. The electron microphotogram

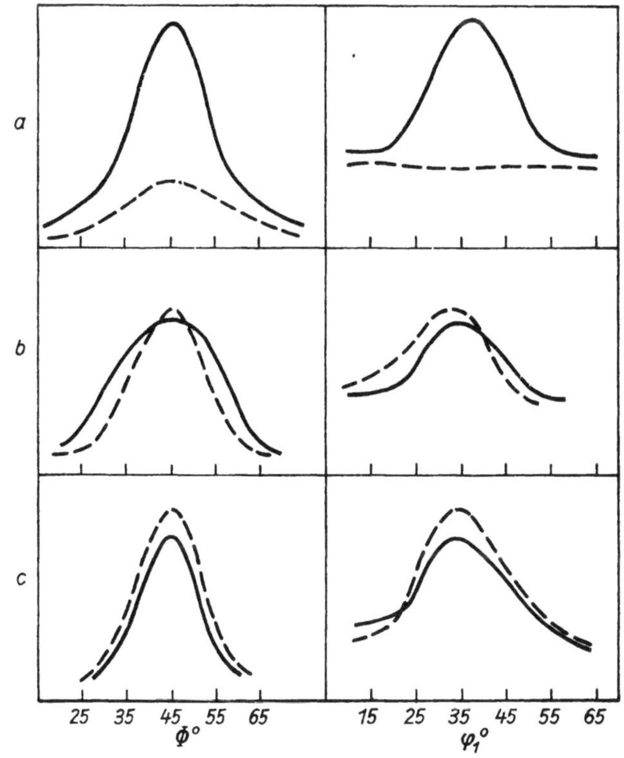

Fig. 4. Crystallite orientation distribution at point "b" perpendicular to and along "a – b" skeleton line for alloy No. 1 (a), alloy No. 2 (b) and alloy No. 3 (c). ——— cold deformed, – – – – annealed

shown in Fig. 3c gives a general idea of alloy structure. Cellular colonies formed by decomposition along grain boundaries (A) can be observed. Also obvious are zones of homogenous $\gamma-\gamma'$ transformation (B) and the formation of fcc phase crystals on intragranular defects with an orientation close to that of the matrix (C).

The second reaction, also a heterogenous one, is the formation of a bcc phase which amounts to 30 to 40 percent. Crystals of this bcc phase nucleate and grow along grain boundaries and slip bands. It is worth noting that in single crystal specimens annealed during up to 100 hrs a transformation of this type has not been observed by VLASOVA (1980). The growth of bcc phase crystals does not cause significant stresses, nor induce the rotation of blocks in the fcc matrix. Consequently, the effect on the texture of the fcc matrix is not noticeable.

It is evident from Figure 4 that the texture of the fcc solid solution becomes more pronounced after the transformations, the peak at point "b" becoming more acute and its half-width reduced. However, the general pattern of the texture is still identical to that of the cold-strained alloy.

It is of interest to compare the texture formation processes in the three alloys. The similarity of the deformation texture in alloys of different composition can be explained by the approximate equality of stacking fault energies, whereas texture changes during annealing are initiated by different processes. In alloy No. 1 the process responsible is recrystallization, with a high probability of nucleation and growth of randomly oriented grains which yield a marked general diffusion of the texture. In the two other alloys recrystallization is suppressed and substituted by the $\gamma-\gamma'$ transformation which yields two fcc phases differing in composition. The transformation follows a strict orientation pattern: the original matrix grain orientation is maintained in the newly formed crystals of fcc phases. In fact, the character and intensity of alloy No. 2 texture is identical to that of deformation texture, whereas in alloy No. 3 the individual features of the deformation texture are augmented. The formation of phases with different fcc structure, more pronounced in alloy No. 3, also affects to a certain degree of the texture formation of the fcc matrix. As concerns the precipitation of hcp and bcc phases, which is a heterogenous reaction and occurs essentially along large-angle boundaries, its effect on texture formation is associated with reducing grain boundary mobility which hinders the growth of randomly oriented grains.

References

BUNGE, H. J.: Texture Analysis in Material Sciences, Mathematical Methods, London 1982
VLASOVA, N. I. et al.: Fizika Metallov i Metallovedenie **49**, 137 (1980)
VLASOVA, N. I. et al.: Izvestia Akademii Nauk SSSR, Metally, No. 1, 86 (1984)

(Received February 26, 1987)

Authors' addresses:

Dr. rer. nat. M. BETZL
Zentralinstitut für Kernforschung der Akademie der Wissenschaften der DDR
PSF 19
Dresden
DDR-8051

Dr. sc. nat. E. N. VLASOVA, Dr. rer. nat. N. B. DJAKONOVA, Dr.-Ing. V. I. MATORIN
Central Research Bardin Institute of the Iron and Steel Industry
2. Baumannskaya
Moscow 107005, USSR

Cryst. Res. Technol.	22	1987	10	1255–1261

U. Steinike, D.-Chr. Uecker, K. Sigrist, W. Plötner, Th. Köhler

Zentralinstitut für physikalische Chemie der Akademie der Wissenschaften der DDR, Berlin, and VEB Elektronik, Gera

Polymorphous Transformation of Quartz Induced by Mechanical Treatment

Because of its strongly damaged structure (partially crystalline) mechanically activated α-quartz is not transformed to β-quartz. Rather, mechanically activated α-quartz is rapidly transformed to the cristobalite modification even at temperatures as low as 1200 °C. The acceleration and temperature shift of the quartz-cristobalite transformation is caused by mechanically induced defects. The activation energy of the transformation is thus reduced by about 50%.

Mechanisch aktivierter α-Quarz zeigt aufgrund der stark gestörten Struktur (partiell-kristallin) nicht mehr die $\alpha-\beta$-Umwandlung. Dafür geht mechanisch aktivierter α-Quarz bereits bei Temperaturen $\geqq$ 1200 °C beschleunigt in die Cristobalit-Modifikation über. Diese beschleunigte und zu tieferen Temperaturen verschobene Quarz-Cristobalit Umwandlung wird durch mechanisch induzierte Defekte hervorgerufen. Die Aktivierungsenergie des Umwandlungsprozesses wird durch die Defekte um ca. 50% reduziert.

1. Introduction

As is generally known, the rate of solid state reactions in a polymorphous conversion is very high (Hedvall effect). Thus, our investigations of some polymorphous conversions of quartz were made mainly in order to obtain information on such solid state reactions, in which quartz is transformed to silicates.

It is especially the quartz-cristobalite transformation that is of great importance for the technological application of respective solid state reactions, since they are accompanied by a considerable change of density (density of quartz 2.6 g/cm³, density of cristobalite 2.2 g/cm³.

As is known from the literature (Sigrist et al.; Senna; Heinicke), the rate and mechanism of modification transformations can be affected by mechanical deformation processes.

The present paper is concerned with the effect of a mechanical treatment on some polymorphous transformations of α-quartz (low-temperature modification).

Without mechanical treatment the polymorphous transformations of the major SiO_2 modifications proceed according to the following scheme (Fenner):

$$\alpha\text{-quartz} \underset{}{\overset{573\ °C}{\rightleftarrows}} \beta\text{-quartz} \overset{870\ °C}{\rightleftarrows} \beta\text{-tridymite} \overset{1470\ °C}{\rightleftarrows} \beta\text{-cristobalite}\,. \tag{1}$$

The present paper is concerned with the following transformations of mechanically activated α-quartz:

$$\alpha\text{-quartz}_{\text{mechanically activated}} \xrightarrow{573\ °C} \beta\text{-quartz}\,, \tag{2}$$

$$\alpha\text{-quartz}_{\text{mechanically activated}} \xrightarrow{\geqq 1200\ °C} \text{cristobalite}\,. \tag{3}$$

2. Experimental

Transformation (2) was investigated with quartz sand from Hohenbocka and Weferlingen (WQS), which was mechanically activated in several types of mills (vibration mill, planetary mill), for 1 to 70 h.

The investigation of transformation (3) was performed on WQS mechanically activated in a vibration mill. The different phases were identified by X-ray diffraction, and in the case of transformation (2) thermoanalytical techniques were employed as well. The kinetics of transformation (3) was studied at 1250 °C and 1350 °C in dependence on the holding time at the respective temperature. The heating time was 5 h. The cristobalite and quartz content were then determined at room temperature by X-ray diffraction (error: ±5%; reflexes used for analysis: d = 4.05 Å, 3.13 Å, 2.84 Å, 2.48 Å (cristobalite) and d = 4.26 Å, 2.45 Å, 2.28 Å, 2.23 Å (quartz)).

Table 1 lists some relevant properties of the quartz samples together with the respective values of Aerosil used for comparison (see below).

Table 1
Characteristic data of the samples used for the quartz-cristobalite transformation

sample	X/% (room temperature)	O m^2/g (room temperature)
WQS 3	84	7.4
WQS 1	86	4.1
WQS 0	99	1
A 200	amorphous	211
A 130	amorphous	130
A 50	amorphous	50

X crystallinity, as determined by X-ray diffraction,
O specific surface (B.E.T. single point technique)
WQS 0, 1, 3 quartz sand from Weferlingen mechanically treated for 0, 1, and 3 h, respectively.

3. Results and discussion

3.1. $\alpha-\beta$-Quartz transformation

KEITH and TUTTLE and SMYKATZ-KLOSS observed differences in the $\alpha-\beta$-transformation temperature for magmatic (573.2 °C) and authigenic quartz (563.6 °C). In the discussion of these differences crystallphysical factors (e.g. defect level) were considered, which may also be relevant for mechanically activated substances. They could be explained, however, by incorporation of impurities. According to FLÖRKE the transformation temperatures are shifted towards lower values (— 560 °C) for very small crystallites (< 0.5 μm), as they can also found upon mechanical treatment. We could not confirm this observation.

Our X-ray diffraction and thermoanalysis results show that mechanically induced defects do not cause a shift of the transformation temperatures in quartz. We found the overall conversion efficiency to decrease with rising defect level; highly damaged α-quartz is not transformed to the high-temperature modification at all. The absence of a conversion effect is considered to result from the strong lattice deformation.

From the point of view of crystal chemistry the $\alpha-\beta$-quartz transformation corresponds to a transformation involving second nearest neighbours. It is a displacive

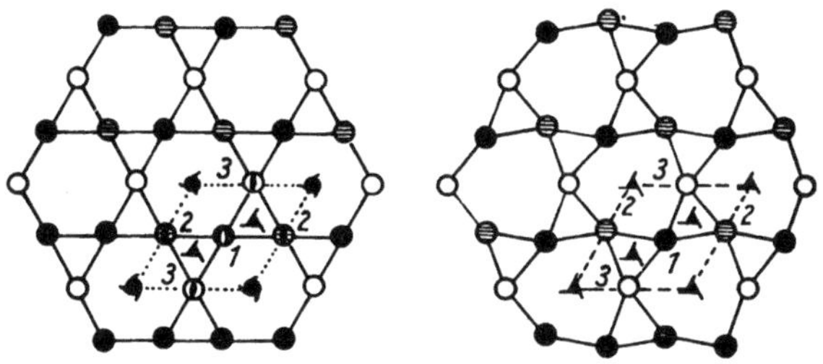

Fig. 1. Schematic representation of the quartz structure (positions of the Si ions given only). a) high-temperature modification; b) low-temperature modification

transformation based on small shifts of the atom positions, i.e. on a lattice deformation. The interconnected tetrahedrons become mutually tilted. The transformation proceeds rapidly; it is reversible. The difference between the α- and the β-modification is schematically shown in Figure 1. In mechanically activated quartz, however, an increasing fraction of the typical hexagonal ring arrangement is damaged, i.e. the quartz is only partially crystalline (STEINIKE et al.).

3.2. Quartz-cristobalite transformation

A significant lowering of the transformation temperature accompanied by a change of the transformation mechanism (see SIGRIST et al. and SENNA) due to mechanical activation is found for the quartz-cristobalite transformation. In the samples cooled to room temperature the cristobalite was identified as α-cristobalite by X-ray diffraction.

The kinetics of cristobalite formation and of quartz transformation (loss of the amount of quartz) can be seen in Figures 2 and 3. Table 2 gives the maximum amounts of cristobalite formed.

X-ray diffraction can only detect the cristobalite and quartz fractions showing good crystallinity. The cristobalite and quartz contents determined by means of a calibration curve do not yield 100% (see Figs. 2 and 3). This difference known from the literature is often explained by using the term "transition phase" (CHAKLADER;

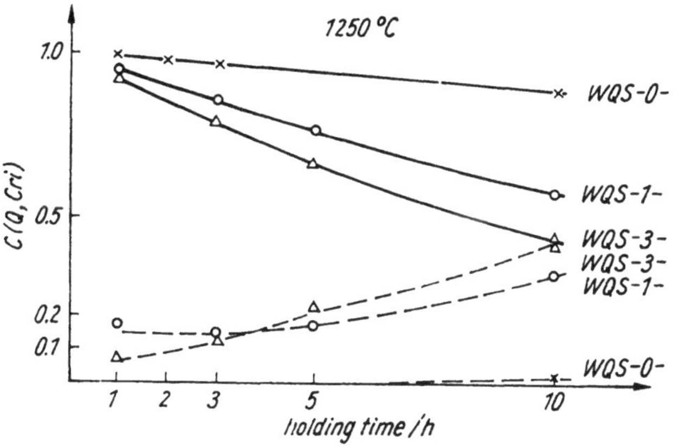

Fig. 2. Dependence of quartz decomposition C_q^* (———) and cristobalite formation C_{cri}^* (– – – –) on holding time t at 1250 °C (WQS 0, 1, 3 quartz sand from Weferlingen mechanically treated for 0, 1, 3 h, resp.) (* mathematically balanced experimental values)

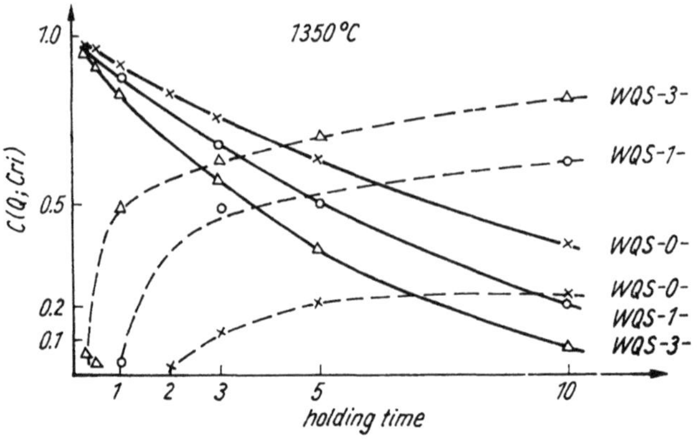

Fig. 3. Dependence of quartz decomposition C_q^* (———) and cristobalite formation C_{cri}^* (– – – –) on holding time t at 1350 °C (* mathematically balanced experimental values)

Table 2
Maximum fractions of cristobalite formation

sample	C_m 1250/%	C_m 1350/%
WQS 3	28	68
WQS 1	20	65
WQS 0	6	53
A 200	3	100
A 130	0	0
A 50	0	0

$C_{m\ 1250}$, $C_{m\ 1350}$ } maximum fractions of cristobalite formed at 1250 °C, and 1350 °C, respectively.

MITRA). This term describes regions about to be transformed, which have not yet developed their final structural order or which are submicroscopically small and thus cannot be detected by X-ray diffraction. From Figs. 2 and 3 it can also be seen that transformation behaviour of mechanically activated quartz is affected by the induced structural defects (e.g. specific surface, cristallinity). The higher the generated defect level (i.e. the longer the time of treatment), the higher the conversion rates and the amount of cristobalite formed.

Like transformation (2) the quartz-cristobalite transformation is connected with a change involving second nearest neighbours. Here, however, the transformation is reconstructive, i.e. the structures have to be completely rearranged involving the breakdown of the Si–O–Si bonds. Such transitions proceed rather slow and within a broad temperature range, as a considerable energy threshold has to be overcome.

The breakdown of the SiO bonds induced by mechanical activation can play a significant role in the observed acceleration and temperature shift of the transformation, as it facilitates the necessary decomposition of the structure.

Assuming 5×10^{14} bond breakages per cm^2 of newly generated surface (HOCHSTRASSER, ANTONIN; STRELETZKIJ, BUTJAGIN), in the quartz samples activated here about 3×10^{19} bond breakages/g should be generated, about $1 \times 10^{19}\ g^{-1}$ of which could be detected as paramagnetic centres (STEINIKE et al.).

Apart from the breakdown of bonds the distortion of the SiO_4 tetrahedrons relative to each other can be involved. Obviously the transformation is additionally facilitated by the partially crystalline structure still containing a fraction of the hexagonal ring arrangement. This assumption is supported by the fact that under the given experimental conditions the formation of cristobalite from amorphous SiO_2 (Aerosil) shows a marked turnover only at temperatures above 1300 °C and for a specific surface $\geqq 200\ m^2\ g^{-1}$ (Tab. 2 and Fig. 4) (SHIUMIZU et al.). Taking the mechanism given by MITRA as a basis,

$$\alpha\text{-quartz} \xrightarrow{k_1} \text{transition phase} \xrightarrow{k_2} \alpha\text{-cristobalite}$$

the constants k_1 and k_2 were determined from the time dependence of the transformation (3). The activation energies A_E of both partial steps can be estimated from the temperature dependence of these constants. The results are given in Table 3.

The mechanical acticvtion results in a reduction of the activation energy of both the transformation quartz → transition phase and the transformation transition phase → cristobalite to half the initial value. The samples milled for 1 h and 3 h can no longer be distinguished. The activation energies of the cristobalite formation step

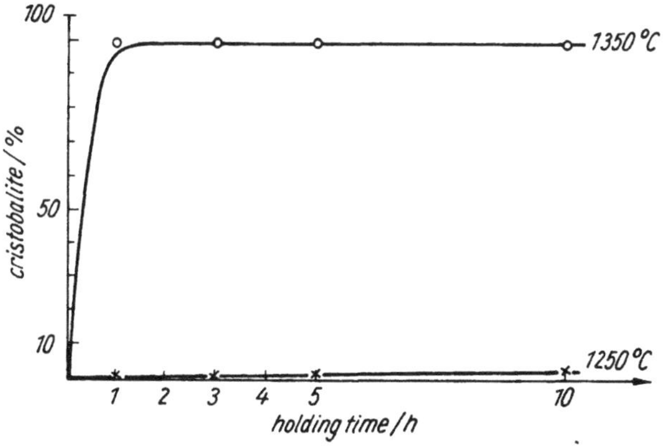

Fig. 4. Dependence of cristobalite formation C_{cri} from Aerosil 200 on holding time t at 1250 °C and 1350 °C

Table 3

Activation energies and rate constants for the mechanically induced quartz-cristobalite transformation

substance	sample	k_1/h^{-1} (α-quartz transition phase)	A_E/kJ mol^{-1} (α-quartz transition phase)	k_2/h^{-1} (transition phase α-cristobalite)	A_E/kJ mol^{-1} (transition phase α-cristobalite)
quartz, before treatment	0/1250	0.0038	405	0.0170	426
	0/1350	0.0278		0.139	
quartz, mechanically activated (1 and 3 h, resp.)	1/1250	0.0166	175	0.553	226
	1/1350	0.040		1.33	
	3/1250	0.0242		0.425	
	3/1350	0.057		1.63	

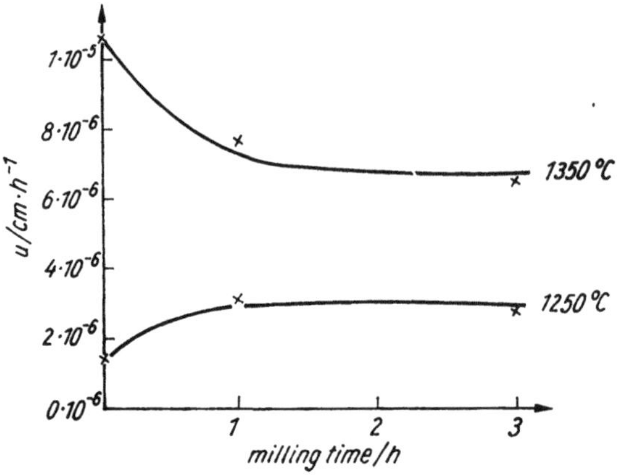

Fig. 5. Dependence of rate u of transformation (3) on milling time t at 1250 °C and 1350 °C

seem to be slightly higher than those of transformation quartz → transition phase. This fact may be due to nucleation processes.

Since the activation energies of both processes are considerably reduced, the mechanically induced defects affect not only the transformation quartz → transition phase, i.e. the decomposition of the quartz lattice, but also cristobalite formation. The effect of the defects should thus still be operative in the supposed transition phase, probably in lattice regions too small to yield diffraction.

Figure 5 shows the dependence of the rate u of transformation (3) on the time t of milling at the temperatures 1250 °C and 1350 °C.

The authors are grateful to Dr. H.-P. Hennig for milling and Dr. U. Kretzschmar and Mr. E. Hennig for determining the specific surface.

References

Chaklader, A. C. D.: J. Amer. Ceram. Soc. **46**, 66 (1963)
Fenner, C. N.: Amer. J. Sci. **36**, 331 (1913)
Flörke, O. W.: Fortschr. Min **44**, 181 (1967)
Heinicke, G.: Tribochemistry, Berlin (DDR) 1983
Hochstrasser, G., Antonini, J. F.: Surface Science **52**, 644, 665 (1972)
Keith, M. L., Tuttle, O. F.: Amer. J. Sci. 208 (1952)
Mitra, S.: Trans. and J. of Brit. Ceram. Soc. **76**, 71 (1977)
Senna, M.: KONA, Powder Science and Technology in Japan **1**, 48 (1983)
Shiumizu, H. et al.: cit. from: C. A. **85**, 134, 156 (1976)
Sigrist, K., Heinicke, G., Steinike, U.: Kristall und Technik **8**, 393 (1973)
Smykatz-Kloss, W.: Contr. Mineral. and Petrol **26**, 20 (1970)
Steinike, U., Müller, B., Ebert, I., Hennig, H. P., Richter-Mendau, J.: Kristall und Technik **14**, 1469 (1979) Sitzungsbericht der Akad. der Wiss. der DDR, Chemie 3 N 150 (1981)
Streletzkij, A. N., Butjagin, P. Ju.: Dokl. Akad. Nauk SSSR **225**, 1118 (1975)

(Received March 4, 1987)

Authors' addresses:

Prof. Dr. habil. U. Steinike, Dipl.-Krist. D.-Chr. Uecker, Dr. K. Sigrist
Akademie der Wissenschaften der DDR, Zentralinstitut für physikalische Chemie
Rudower Chaussee 5
Berlin
DDR-1199

Dipl.-Ing. W. Plötner, Dipl.-Ing. Th. Köhler
VEB Elektronik Gera
Parkstraße 3
Gera
DDR-6500

Book Review

A. J. Pertsin, A. L. Kitaigorodsky

The Atom-Atom Potential Method. Applications to Organic Molecular Solids

Springer Series in Chemical Physics Vol. 43

Springer-Verlag Berlin, Heidelberg, New York, London, Paris, Tokyo 1987, 397 Seiten
ISBN 3-540-16246-1 Springer-Verlag Berlin, Heidelberg, New York, ISBN 0-387-16246-1
Springer-Verlag New York, Berlin, Heidelberg, DM 140,–

Das Buch ist eine moderne Übersicht über die Nutzung von Atom-Atom-Potentialen zur Berechnung und Interpretation der Eigenschaften von organischen Molekülkristallen. Empirische Potentiale — historisch mit dem Namen Kitaigorodsky verbunden — werden in den letzten Jahren zunehmend durch quantenchemische ab-initio-Potentiale verbessert bzw. ersetzt. Der aktuelle Stand dieser Entwicklung wird ausführlich dargestellt.

Nach einer Einführung werden in Kapitel 2 die gegenwärtigen Möglichkeiten der nichtempirischen Berechnung der zwischenmolekularen Kräfte zwischen organischen Molekülen (Supermolekülansatz, Störungstheorie) behandelt und an ausgewählten Beispielen aus der Literatur recht ausführlich erläutert. Dieses Kapitel nimmt keinen Bezug auf die im Titel ausgewiesene Anwendungsrichtung, sondern könnte gleichermaßen einem Buch über die Anwendung der Methode auf Flüssigkeiten, Gase bzw. biologisch interessante Systeme voranstehen. In Kapitel 3 werden die Atom-Atom-Potential-Methode selbst und die verschiedenen Potentialansätze eingeführt. Es erfolgt eine ausführliche Diskussion über die Möglichkeiten der Auswahl und der Justierung der Potentialparameter. Zahlreiche Parametersätze aus der Literatur wurden zusammengestellt. Auch Wechselwirkungen mit Schwefel, Selen, Sauerstoff, Stickstoff und Halogenen sind berücksichtigt.

Die Autoren legen Wert auf die physikalische Interpretation der bei der Zerlegung der Wechselwirkungsenergie auftretenden Energieterme, insbesondere aber auf die Vergleichbarkeit dieser Terme bei verschiedenen Zerlegungen.

In weiteren Kapiteln erfolgt die Anwendung der Methode auf die Gitterstatik, die Gitterdynamik und die Thermodynamik organischer Molekülkristalle. Ein breites Spektrum von Kristalleigenschaften wurde behandelt, die Grenzen und Möglichkeiten der Verwendung von Atom-Atom-Potentialen zur Berechnung dieser Eigenschaften werden diskutiert. Genannt seien etwa die Berechnung von Kohäsionsenergien, Gleichgewichtskristallstrukturen (auch von Polymerkristallen), Kraftkonstanten, thermodynamischen Funktionen und damit die Interpretation von Röntgenstrukturdaten, Phononenspektren, polymorphen Übergängen usw. Das letzte Kapitel behandelt die Berechnung der Gitterenergie in Kristallen mit Strukturdefekten.

Das Buch hat ausgesprochen interdisziplinären Charakter. Es spricht Festkörperphysiker und -chemiker, Physikochemiker, Polymerchemiker, Kristallographen und Quantenchemiker gleichermaßen an. Die relativ ausführlichen konkreten Beispiele und die zahlreichen, sorgfältig ausgewählten Literaturzitate seien hervorgehoben.

J. Reinhold

Cryst. Res. Technol.	22	1987	10	1263–1269

T. FIGIELSKI

Institute of Physics, Polish Academy of Sciences, Warsaw

Antistructure Disorder and its Relation to Dislocations in III–V Semiconductors[1])

Intimate relationship between antistructure defects and dislocations occurs in GaAs which manifests itself as: (i) appearing of spatial correlation between grown-in dislocations and the EL2 defects (component of the latter are As_{Ga} antisites), (ii) similar suppression of the concentration of EL2 and dislocation density due to the doping with donor impurities, (iii) generation of As_{Ga} antisites during the plastic deformation of a crystal. All these effects can be understood in terms of the dislocation-mediated generation of As_{Ga} antisites via absorption of As_i interstitials at dislocation jogs. Large anion (cation) precipitates appearing at dislocations are pointed out to be important for both antisite and dislocation generation under certain conditions.

1. Introduction

In a two-component semiconductor compound, AB, some of A atoms may be misplaced at the sites of B atoms, and vice versa, giving rise to the antistructure disorder. This kind of disorder has been considered in textbooks of physical chemistry of solids for a long time together with two other symmetrical types of disorder, i.e. Schottky and interstitial disorders, and their three hybrids (see e.g. KRÖGER). But it is the last few years when the antistructure disorder was experimentally verified in III–V compounds and its importance for electronic properties of these materials directly demonstrated. Crucial points were the identification of some anion antisite defects in III–V's with EPR technique. Presently are known: P_{Ga}^{+} in GaP (KAUFMANN et al.), As_{Ga}^{+} in GaAs (WAGNER et al.), and P_{In}^{+} in InP (KENNEDY, WILSEY). A real explosion of researches on this field arrived when it was recognized that the native defect responsible for semi-insulating properties of undoped GaAs, called EL2, is directly related with As_{Ga} antisite. Up to date a huge amount of data were collected for EL2 and As_{Ga} defects in GaAs (see e.g. MARTIN, MAKRAM-EBEID). Just the latter system will be considered in this paper which is believed to be also representative for other III–V's.

2. EL2 problem in GaAs

EL2 is the dominant native donor in melt-grown GaAs which introduces into the band gap two energy levels, the upper one being located at the middle of the gap. This midgap level controls the Fermi level position in undoped LEC-grown GaAs crystals making them semi-insulating which is of great technological importance. Characteristic

---

[1]) Paper presented at 21. Jahrestagung der Vereinigung für Kristallographie, Leipzig, DDR, February 25–27, 1987.

peculiarity of EL2 is its ability to be transformed at low temperature into a metastable state under illumination with $\lambda = 1.06$ μm wavelenght radiation.

It is now well documented that As_{Ga} antisite is at least one of the components of the EL2 defect although the detailed structure of EL2 is still unknown. Three microscopic models of EL2 are considered at present. The first, based on the symmetry arguments, identifies EL2 with the isolated As_{Ga} defect (KAMIŃSKA et al.). The second, based primarily on the results of optically detected ENDOR measurements claims that in addition to As_{Ga} antisite also As_i interstitial is involved in EL2 (MEYER et al.). The latter model is in conflict with the piezospectroscopic results which point out a high symmetry of the defect. Recently, a compromise model was proposed according to which EL2 is composed of the $(As_{Ga})_2$ dimer, whose the two constituents are the second neighbours along a ⟨110⟩ axis, plus As_i interstitial in the position symmetric with respect to the both (FIGIELSKI 1986). Such a complex has C_{2v} symmetry, and reconciles the two apparently opposed findings: involvement of As_i interstitial in EL2, and appearance of a high symmetry of this centre (FIGIELSKI, WOSIŃSKI).

It was very early recognized that a close relationship between EL2 or As_{Ga} and dislocations appears in GaAs. It manifests itself in several ways and seems to be of more fundamental nature than simple gettering of defects to dislocations. The aim of this paper is to reveal and discuss the nature of that relationship.

3. Grown-in dislocations and EL2

Typically, dislocations in LEC-grown GaAs crystals exhibit overall W-shaped, distribution across the wafer diameter, at least in the seed part of the crystal (Fig. 1), with superimposed cellular structure. The same features have been often observed in the distribution of EL2 centres as determined from the near infrared (IR) absorption at $\lambda \approx 1$ μm that is associated with EL2 (see e.g. FIGIELSKI 1987 and literature therein). In the recent work ALT and PACKEISER compared the IR absorption maps and transmission X-ray topographs taken from the same wafers cut from the seed end of GaAs crystals grown by LEC method. They established qualitative one-to-one spatial correlation between EL2 distribution and grown-in dislocation networks

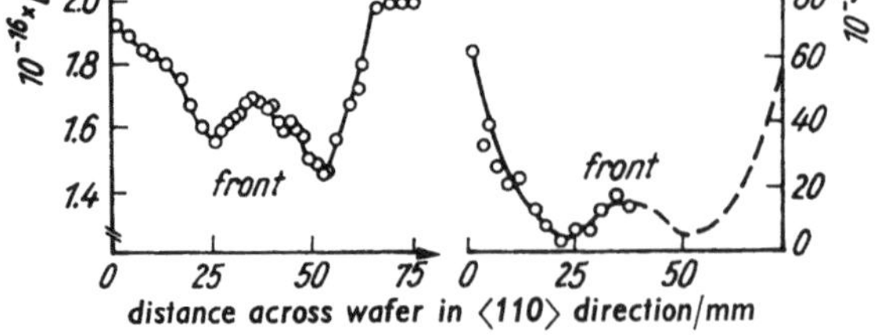

Fig. 1. EL2 concentration and etch pit density (EPD) across diameter of GaAs crystal grown by LEC method from near-stoichiometric melt, according to HOLMES et al.

even on a microscopic scale down to 100 μm resolution. It was independently verified that the fine structure observed on the IR absorption maps exhibit persistent photoquenching behaviour characteristic of EL2 metastable state (SKOLNICK et al.). At closer inspection of the IR absorption topographs it has been inferred, from the lack of denuded zones around dislocations, that the EL2 defects likely are generated at the dislocations and not collected only. Moreover, after high temperature annealing uniformity of EL2 distribution is improved.

It is worth to notice here that better resolution in distribution of EL2, at least down to 10 μm, offers Scanning DLTS technique. First attempts of applying this technique to the present problem were in fact successfully done by BREITENSTEIN and WOSIŃSKI, and WOSIŃSKI and BREITENSTEIN.

Another type of correlation between grown-in dislocations and EL2 was revealed in MIT group where one studied an effect of crystal doping with electrically active impurities (LAGOWSKI, GATOS). Concentrations of both the dislocations and EL2 are suppressed in a similar way when donor concentration exceed a value of 5×10^{16} cm^{-3} (Fig. 2). LAGOWSKI and GATOS, to explain this dependence, claims that both As_{Ga} antisites and V_{As} vacancies are generated in a crystal from originally present V_{Ga} vacancies through the reaction:

$$V_{Ga}^{--} + As_{As} \rightleftarrows As_{Ga}^{++} + V_{As}^{+} + 4e^{-} , \qquad (1)$$

which is controlled by the Fermi level position in the crystal due to appearance of free electrons on the right hand side of eq. (1). Their principal assumption is that the dislocations are generated by condensation of excess vacancies into dislocation loops. Thus both EL2 and dislocations would have their common origin in the process (1) controlled by the Fermi level. This explanation, however, hardly can be accepted since glide and not climb is the most probable origin of dislocations in melt-grown GaAs. The principal argument for that is the W-shaped distribution of dislocation density across the crystal diameter which is theoretically predicted by consideration of the thermal stress involved during cooling the crystal grown from the melt (JORDAN).

According to the present writer opinion a dislocation-mediated generation of As_{Ga} antisites (or antisite-related defects) is primarily responsible for the observed correlations. Obviously, an excess of arsenic is necessary for generation of As_{Ga} antisites, and it actually was proved to favour the production of EL2. A large excess of As_i interstitials is expected to arise in GaAs crystals growing from As-rich melt just after their solidification. During the post-solidification cooling mobile interstitials move to sinks such as the surface and dislocations. Point defects adjoined to dislocation core migrate along the core until they annihilate at dislocation jogs giving rise to the dislocation climb (Fig. 3).

Usually, jogs represent constricted segments of the split dislocations and may appear of various heights. In GaAs two different atom species, i.e. As and Ga, are required to be absorbed at a regular jog of unit height to reproduce it at an adjacent position along the dislocation line. However, under conditions of supersaturation with As_i interstitials only, the predominant process will be the formation of irregular jogs by absorption of two As_i atoms at the same jog (FIGIELSKI 1985). If the dislocation involved in the process is of the 60°-type then after its glide together with the jog a misplaces atom is left behind: As at the site of Ga, i.e. the As_{Ga} antisite defect. In a similar way also complexes of As_{Ga} defects can be produced due to absorption of more than two As_i atoms at the same jog, e.g. when the jog is of more than unit height. This is the mechanism of the generation of As_{Ga} antisites, and their complexes, which is conditioned both upon excess arsenic and dislocation glide.

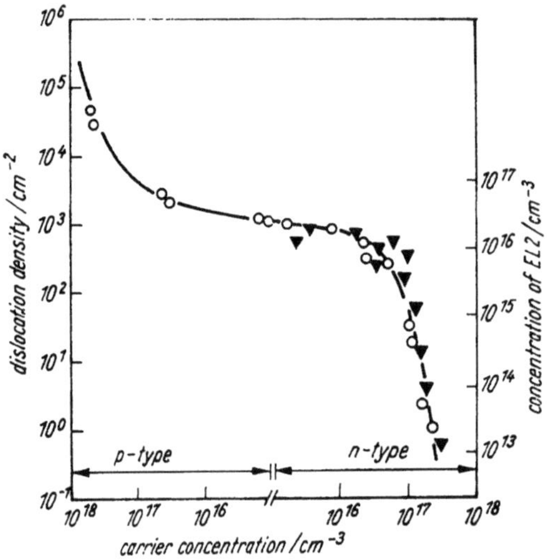

Fig. 2. Effect of doping with electrically active impurities on EL2 (▼) and dislocation density (○) in GaAs, according to LAGOWSKI and GATOS

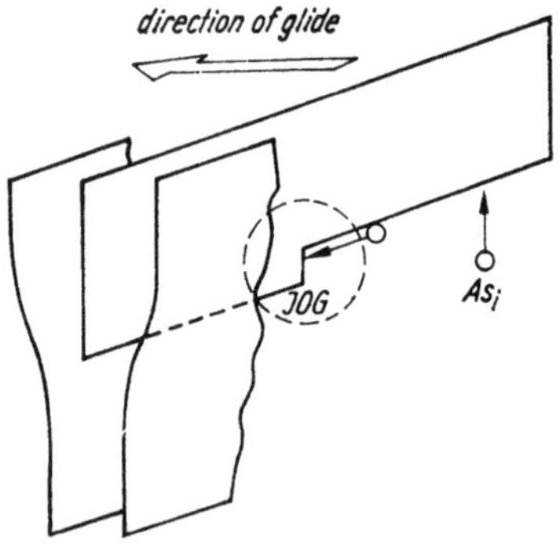

Fig. 3. Absorption of two As$_i$ atoms at dislocation jog

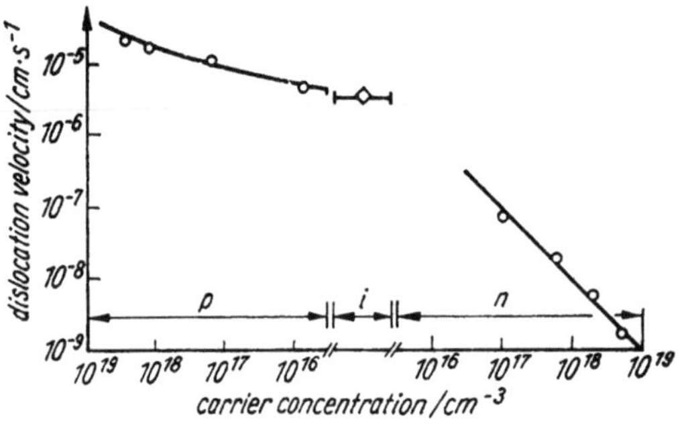

Fig. 4. Velocity of β-dislocations in GaAs as a function of doping, according to MAEDA

It may be assumed that predominantly β-dislocations are involved in the mechanism of the As_{Ga} antisite generation. If the dislocations in GaAs belong to the glide set, as commonly accepted, then the core atoms of β-dislocation are cations which favours the absorption of anion atoms (As) at the core. It is now well documented that dislocation glide mobility in covalent semiconductors, which is controlled by the lateral motion of kinks, may be affected by doping with electrically active impurities through a change of the Fermi level of the crystal. This effect usually is interpreted in terms of the dependence of kink density and kink migration energy on the Fermi level (see e.g. HAASEN).

Fig. 4 presents the observed velocity of β-dislocations in dependence on doping (MAEDA), that is seen to display the same characteristic features as the concentration of EL2 and dislocation density do, i.e. a rapid decrease at $n > 5 \times 10^{16}$ cm^{-3} and slow increase at $p > 5 \times 10^{16}$ cm^{-3}. This gives a strong argument in favour of the dislocation-mediated generation of EL2 and is an alternative explanation of the dependences presented in Figure 2.

4. Effect of plastic deformation and non-stoichiometry

A separate phenomenon, which is being discussed now, is the generation of As_{Ga} antisites during low-temperature plastic deformation of GaAs crystals. The deformation is usually performed at a temperature 350—400 °C and it gives rise to an increase in the amplitude of EPR spectrum characteristic of the As_{Ga}^{+} paramagnetic centre (Wosiński; Weber et al.). This increase seems to be caused by a real enhancement of the concentration of As_{Ga} antisites and not only by the change of their charge state, although the latter cannot be completely excluded so as a high concentration of acceptors with the energy levels in the lower half of the band gap is generated in GaAs due to the deformation (Suezawa, Sumino; Gerthsen, Haasen).

Assuming that during the plastic deformation As_{Ga} defects are generated by interaction of As_i interstitials with moving dislocations, one has to indicate a source of the interstitials since it is unlikely that isolated interstitials were available in the crystal under post-growth condition. Presumably, such sources are large As precipitates (a few hundred Å of diameter) usually located along dislocation lines, which have been observed in melt-grown GaAs for some time (Bartels et al.; Cullis et al.; Gant et al.). It is likely that at the deformation temperature some As_i are released from the precipitates, in accordance with the observation that anion precipitates in GaAs disappear at 300 °C and reappear at lower temperature but not necessarily at the same positions as before heating. It is supposed that mobile As_i interstitials, released from the precipitate particles, interact with gliding dislocations and undergo transformation into the As_{Ga} defects, e.g. through the irregular jog formation.

It is worth to notice here that essential precipitation of anion or cation species has to be also assumed in order to explain why semi-insulating GaAs crystals can be obtained by LEC growth from the melt in which the arsenic fraction ranges from 0.475 to 0.535. Such large nonstoichiometry of the melt could not be accomodated in the crystal by simple native defects alone since the concentration of the dominant defect (EL2) is at least three orders of magnitude too low.

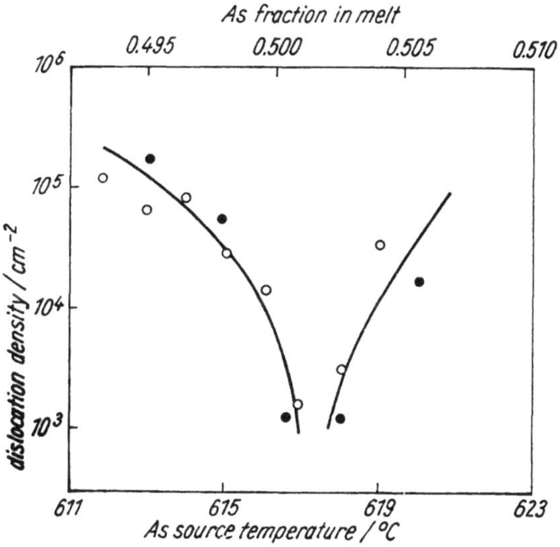

Fig. 5. Dislocation density in dependence on melt stoichiometry for lightly doped n-type (●) and p-type (○) GaAs, according to Lagowski et al.

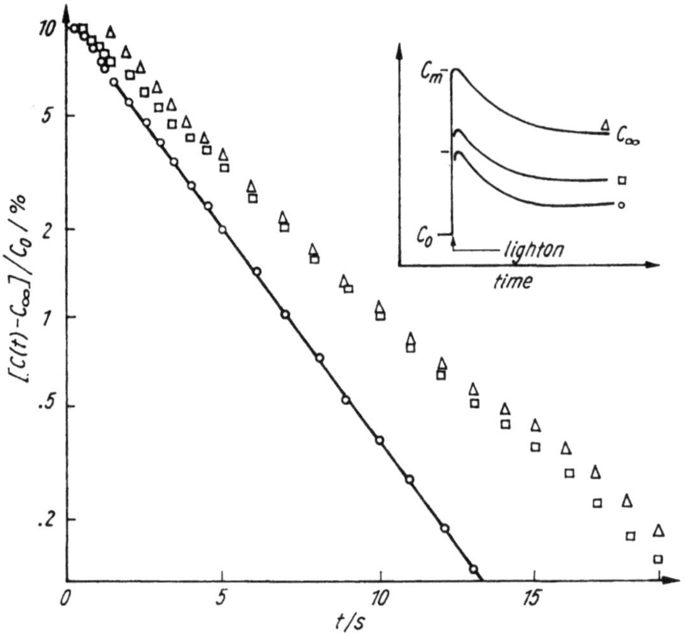

Fig. 6. Transient of photocapacitance quenching at 77 K under $\lambda = 1.06$ μm illumination, due to transformation of EL2 into metastable state, for undeformed (○) and plastically deformed ($\varepsilon = 0.5\%$, □; $\varepsilon = 2.8\%$, △) n-type GaAs, according to WOSIŃSKI and FIGIELSKI

Another manifestation of the action of anion (cation) precipitates in GaAs is possibly the appearing of a distinctive dependence of grown-in dislocation density on the melt stoichiometry, as shown in Figure 5. At the same time the concentration of EL2 rises monotoneously with increasing As fraction in the melt. This result was obtained by MIT group for horizontal Bridgman growth of GaAs with the melt stoichiometry varied by changing the temperature of the arsenic source (LAGOWSKI et al.). A rapid increase of the dislocation density with the deviation from melt stoichiometry may be assigned to the punch-in effect of the precipitate particles which act as internal sources of dislocation loops.

To complete this section some recent results concerning the plastically deformed GaAs (WOSIŃSKI, FIGIELSKI) should be mentioned. It has been definitely established that the deformation has a little effect on the concentration of EL2 itself whereas it primarily increases the concentration of the electron trap called EL6 which is characteristic of the thermal ionisation enthalpy 0.35 eV. For several reasons this EL6 has been tentatively assigned to the isolated As_{Ga} antisite. It is also noteworthy that in consequence of the plastic deformation the kinetics of transformation of EL2 into the metastable state, under illumination with $\lambda = 1.06$ μm radiation, becomes slower and non-exponential (Fig. 6), which may be explained as due to an interaction between the EL2 and EL6 defects.

The results discussed in this paper strongly suggest an intimate relationship between the antistructure defects and dislocations in GaAs. Although our speculations are able to explain some of the experimental results, in general, this relationship which seems to be of very fundamental origin contains many puzzles. Perhaps new findings are necessary to comprehense all the phenomena involved in the alliance between the antistructure disorder and topological (dislocation) disorder appearing in III–V semiconductors.

References

ALT, H. Ch., PACKEISER, G.: J. Appl. Physics **60**, 2954 (1986)
BARTELS, F., CLEMENS, H. J., MÖNCH, W.: Physica **117B** a. **118B**, 801 (1983)
BREITENSTEIN, O., WOSINSKI, T.: phys. stat. sol. (a) 77, K107 (1983)
CULLIS, A. G., AUGUSTUS, P. D., STIRLAND, D. J.: J. Appl. Physics **51**, 2556 (1980)
FIGIELSKI, T.: Appl. Physics **A36**, 217 (1985)
FIGIELSKI, T.: Material Science Forum **10—12**, 341 (1986)
FIGIELSKI, T.: Acta Phys. Polon. (1987) in print
FIGIELSKI, T., WOSINSKI, T.: Phys. Rev. B, **36**, 1269 (1987)
GANT, H., KOENDERS, L., BARTELS, F., MÖNCH, W.: Appl. Phys. Lett. **43**, 1032 (1983)
GERTHSEN, D., HAASEN, P.: Acta Phys. Polon. **A69**, 415 (1986)
HAASEN, P.: J. de Phys. **40**, C6—111 (1979)
HOLMES, D. E., CHEN, R. T., YANG, J.: Appl. Phys. Lett. **42**, 419 (1983)
JORDAN, A. S.: J. Cryst. Growth **49**, 631 (1980)
KAMINSKA, M., SKOWRONSKI, M., KUSZKO, W.: Phys. Rev. Lett. **55**, 2204 (1985)
KAUFMANN, U., SCHNEIDER, J., RÄUBER, A.: Appl. Phys. Lett. **29**, 312 (1976)
KENNEDY, T. A., WILSEY, N. D.: Appl. Phys. Lett. **44**, 1090 (1984)
KRÖGER, F. A.: The Chemistry of Imperfect Crystals, 1964
LAGOWSKI, J., GATOS, H. C.: Proc. 13th Intern. Conf. Defects in Semiconductors, Coronado 1984, p. 73
LAGOWSKI, J., GATOS, H. C., AOYAMA, T., LIN, D. G.: Appl. Phys. Lett. **45**, 680 (1984)
MAEDA, K.: in: Dislocations in Solids, edited by H. SUZUKI, 1985, p. 425
MARTIN, G. M., MAKRAM-EBEID, S.: Deep Centres in Semiconductors, edited by S. T. PANTELIDES, 1986, p. 399
MEYER, B. K., HOFMANN, D. M., SPAETH, J.-M.: Material Science Forum **10—12**, 311 (1986)
SKOLNICK, M. S., REED, L. J., PITT, A. D.: Appl. Phys. Lett. **44**, 447 (1984)
SUEZAWA, M., SUMINO, K.: Jap. J. Appl. Physics **25**, 553 (1986)
WAGNER, R. J., KREBS, J. J., STAUSS, G. H., WHITE, A. M.: Solid State Commun. **36**, 15 (1980)
WEBER, E. R., ENNEN, H., KAUFMANN, U., WINDSCHEIF, J., SCHNEIDER, J., WOSINSKI, T.: J. Appl. Physics **53**, 6140 (1982)
WOSINSKI, T.: phys. stat. sol. (a) **60**, K149 (1980)
WOSINSKI, T., BREITENSTEIN, O.: phys. stat. sol. (a) **96**, 311 (1986)
WOSINSKI, T., FIGIELSKI, T.: Solid State Commun. **63**, 885 (1987)

(Received March 25, 1987)

Author's address:

Prof. T. FIGIELSKI
Instytut Fizyki PAN
02-668 Warszawa
Al. Lotników 32/46

Book Review

E. S. MASHKOVA, V. A. MOLCHANOV

Medium-Energy Ion Reflection from Solids

Modern problems in condensed matter sciences vol. 11

North-Holland Amsterdam-Oxford-New York-Tokyo 1985, 434 S., 223 Abb. Preis: US $ 152.00/ Dfl. 380.00 ISBN 0-444-86945-X. In the USA/Canada the book is available from Elsevier Science Publ., New York

Diese bisher 16 Bände umfassende Serie erscheint seit 1982 und stellt ein Gemeinschaftsunternehmen überwiegend sowjetischer und US-amerikanischer Wissenschaftler dar. Die jeweiligen Bände wurden von „North-Holland" und „Nauka" etwa zeitgleich in englischer und russischer Sprache herausgegeben. Im Band 11 kommen zwei profilierte sowjetische Wissenschaftler zu Wort, die seit Jahrzehnten experimentell und theoretisch auf dem Gebiet der Wechselwirkung von Ionen mittlerer Energie (ca. 1—100 keV) mit poly- und einkristallinen Materialien arbeiten. Historischer Ausgangspunkt dieser Untersuchungen waren Fragen der Plasma-Wand Wechselwirkung, die für alle Fusionsexperimente von grundsätzlicher Bedeutung sind. (Das Buch ist L. A. ARZIMOVICH, dem Pionier der sowjetischen Fusionsforschung, gewidmet.)

In der seitdem vergangenen Zeit hat sich die Anwendungsbreite des im Titel genannten Forschungsgebietes durch die Entwicklung der Ionenimplantation and analytischer Ionenstrahltechniken wesentlich erweitert. Der vorliegende Band stellt eine erweiterte und überarbeitete Fassung eines von den Autoren bereits im Jahre 1980 im Verlag „Atomisdat" (Moskau) herausgegebenen Buches dar.

In den ersten drei Kapiteln der jetzigen Ausgabe wird die Theorie der Ionenstreuung im betrachteten Energiegebiet am ungeordneten und geordneten Festkörper dargestellt. Dabei erfolgt eine systematische Darlegung der Streuung am Einzelatom, am Atompaar und an der Atomfläche, um so schließlich die Streuung am Kristall beschrieben zu können.

Es werden dabei sowohl analytische Modelle als auch Computersimulationen berücksichtigt. Kapitel 4 stellt relativ ausführlich Experimentieranlagen (Ionenbeschleuniger, Streukammern und Detektorsysteme einschl. Datenverarbeitung) aus verschiedenen Laboratorien unterschiedlicher Länder vor. Dieses Kapitel ist gut geeignet, um sich schnell einen Überblick über die z. Z. verwendete Meßtechnik zu verschaffen. In den folgenden Kapiteln (5—10) werden experimentelle Ergebnisse zur Winkel- und Energieverteilung der an amorphen, polykristallinen und einkristallinen Materialien gestreuten Ionen und der erzeugten Rückstoßatome vorgestellt, diskutiert und mit theoretischen Modellen verglichen. Weitere Parameter sind Einschußwinkel, Inzidenzenergie und Targettemperatur.

Die Forschungsergebnisse der Autoren nehmen hier einen wichtigen Platz ein, allerdings nicht zu Lasten einer umfassenden Literaturauswertung. Relativ knapp werden im Kapitel 11 Anwendungsaspekte behandelt. Die Autoren verweisen auf schon publizierte Monographien und Konferenzbände zu dieser Thematik. Im wesentlichen beschränken sie sich auf die Darstellung von Untersuchungen an relaxierten Oberflächenatomen kristallinen Materials und von oberflächenanalytischen Arbeiten zur Spektroskopie von Rückstoßteilchen.

Das Buch kann all denjenigen von Nutzen sein, die sich experimentell oder theoretisch mit der Wechselwirkung von Ionen mittlerer Energie mit Festkörpern beschäftigen bzw. einarbeiten wollen.

V. GEIST

Cryst. Res. Technol.	22	1987	10	1271–1275

B. J. MEHTA, G. R. SIDPARA

Solid State Physics and Materials Science Laboratory, Department of Physics, Saurashtra University Rajkot, India

Gross Dissolution of Natural Rhombohedral Calcite Crystal in Nitric Acid

Gross dissolution of natural calcite crystal in concentrated nitric acid give interesting results. The gross dissolution is maximum at 6 M concentration of HNO_3. The activation energy increases as the concentration of HNO_3 increases. Similarly pre-exponential factor is also increases as concentration increases. The combined effect of activation energy and pre-exponential factor gives the peak at 6 M concentration of nitric acid.

Introduction

It is known that calcite is very reactive with acids. The different concentrations of acids such as HNO_3, HCl, CH_3COOH etc. produces dislocation etch pits and different etch pit morphology on cleavage faces of calcite is studied (MEHTA; MEHTA & PANDYA). Many authors (SMIRNOV, URUSOVSKAYA; KONNEKE, LACMANN; HARTMAN et al.) have investigated the gross dissolution of sphere shaped crystals like BaF_2, copper and gadolinium gallium garnet. In this paper the gross dissolution of natural rhombohedral calcite crystals is presented.

Experimental

Transparent rhombohedral calcite crystal from Gondal (Dist. Rajkot, India) are selected for gross dissolution study. Crystals of size about 2 to 3 cm^3 were etched in steady etchant for desire time of etching at constant temperature and then washed in running distilled water and dried by hot air. The crystals were weighed before and after etching by a chemical balance of accuracy 0.1 mg. The dimensions of crystal were measured before and after etching with traveling microscope having accuracy of 0.001 cm. Surface area and volume were calculated from the dimensions.

Results

Large number of rhombohedral crystals were etched in nitric acid concentration ranging from 1.5 to 7.5 M at various temperatures. The weight loss, volume loss, and surface area loss per unit area were measured for large number of calcite crystals and Table 1 was prepared for different temperatures.

Figure 1 shows the graph of weight loss per unit area versus concentration of nitric acid for different temperatures. The graph shows peak at perticular concentration for all the values of temperatures. The gross dissolution decreases as the concentration decreases or increases from 6 M of HNO_3. It should be mentioned that the concen-

Table 1

Values of weight loss or volume loss or surface area loss per unit area by different molarity of nitric acid at different temperatures in gross dissolution of rhombohedral calcite crystal

Concentration C/M	temperature t/°C	W/A gm · cm^{-2}	V/A cm	A/A
1.5	30	0.005	0.003	0.009
3.6		0.012	0.005	0.015
4.5		0.015	0.006	0.018
6.0		0.017	0.007	0.020
7.5		0.012	0.005	0.017
1.5	40	0.007	0.004	0.014
3.0		0.018	0.008	0.023
4.5		0.022	0.011	0.032
6.0		0.027	0.013	0.037
7.5		0.022	0.011	0.029
1.5	50	0.009	0.005	0.018
3.0		0.022	0.011	0.031
4.5		0.032	0.014	0.039
6.0		0.040	0.016	0.052
7.5		0.036	0.014	0.049

Etching time 5 seconds

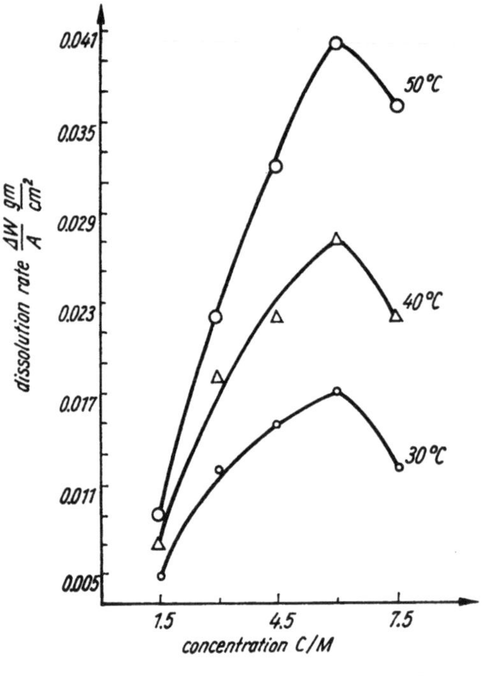

Fig. 1. Plots of $\Delta W/A$ versus concentration of nitric acid at various temperatures

tration versus C/n^2 shows the same behaviour as shown in Figure 1, where n is viscosity of etchant. It is found that there is no change in rhombohedral shape after etching. Figure 2 shows the graph of log of weight loss per unit area versus reciprocal of absolute temperatures for different concentration of nitric acid. The graph is straight

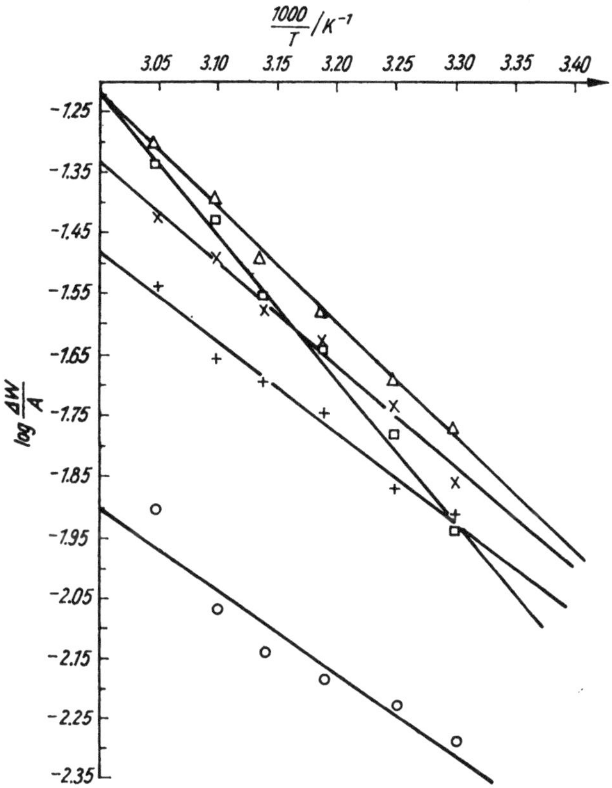

Fig. 2. Plots of log $\Delta W/A$ versus $1/T$ for different nitric acid concentrations: ○ 1.5 M, + 3.0 M, × 4.5 M, △ 6.0 M, □ 7.5 M

Table 2

Values of activation energy and pre-exponential factor for different molar nitric acid

Observation No.	Concentration C/M	activation energy E_a/eV	pre-exponential factor A gm/cm² · s
1	1.5	0.27	$1.51 \cdot 10^2$
2	3.0	0.30	$1.05 \cdot 10^3$
3	4.5	0.34	$0.63 \cdot 10^4$
4	6.0	0.38	$0.33 \cdot 10^5$
5	7.5	0.48	$1.12 \cdot 10^6$

line and shows negative slope. For different concentration activation energy and pre-exponential factor was calculated and Table 2 is prepared. If we draw graph of logarithium of weight loss per surface area or volume loss per surface area or surface area loss per surface area versus reciprocal of absolute temperature for different concentration, the graph is straight line with negative slopes and activation energy was found to be nearly the same in all the methods. It was also found that there is no change of order in pre-exponential factor which is founded by above three methods. Activation energy and pre-exponential factor both increases with concentration and their combined effect gives rise to peak value at concentration of 6 M HNO_3.

Discussion

The activation energy was found between 0.27 to 0.48 eV in the concentration range from 1.5 to 7.5 M. This suggests that the reaction is diffusion controlled reaction. In diffusion process H^+, NO_3^{-1}, and OH^- ions adsorb at the calcite surface, and they make complexes like calcium hydroxide and calcium nitrate. The number density of above complexes depends upon the concentration of etchant. These complexes then turn to reaction products like $Ca(NO_3)_2$ and $Ca(OH)_2$, the former dissolve faster in etchant whereas $Ca(OH)_2$ dissolves slower. The production of $Ca(OH)_2$ produces heat. From the Table 2 it is found that the activation energy increases with increase in concentration. This can be understood by assuming formation of above two complexes one is calcium hydroxide and other is calcium nitrate. As concentration increases the formation of calcium nitrate complex is more prominant and as concentration decreases the formation of calcium hydroxide complex becomes more prominant. The formation of complex calcium hydroxide results into product $Ca(OH)_2$, and the process is exothermic. This will lead to local increase of heat at surfaces which make the decrease in apparent activation energy. This phenomenon is true at very low concentration where the dislocation etch pits are formed. The activation energy is found from lateral development of size of pits and reciprocal of absolute temperature. The activation energy is to be found 0.15 and 0.07 eV at concentration 0.225 and 0.03 M of HNO_3, respectively. From the Table 2, it is found that activation energy increases with concentration. To find a relation between activation energy and concentration graph of activation energy versus square of concentration is drown and shown in Figure 3. The graph is nearly a straight line, and it shows the slope is equal to 0.004.

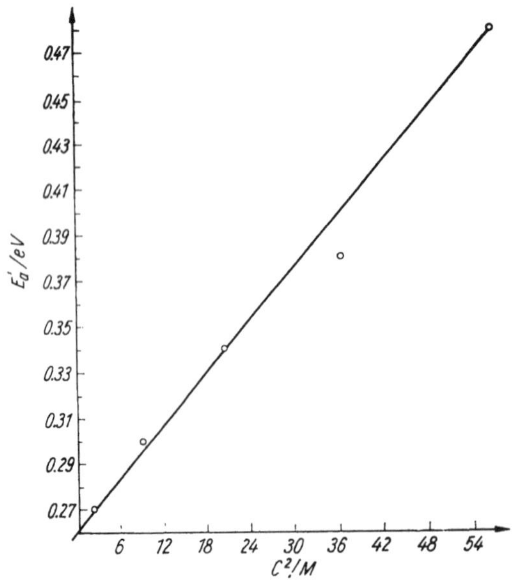

Fig. 3. Plot of activation energy in eV versus square of concentration of etchant in molarity of nitric acid

So for gross dissolution of calcite crystal in HNO_3 acid the equation representing the activation energy and concentration is nearly

$$E_a = 0.004C^2 + 0.26\,.$$

It is also found that the pre-exponential factor A also increases with the increase in concentration. It is known that A depends upon the steric factor and collision frequency. As concentration increases the viscosity of etchant also increases. The viscosity of etchant resist the product elements and complexes come to closer into solution and so complexes again act on the effected surface or they stay more time at the surfaces. This will increase the collision frequency and hence pre-exponential factor. Viscosity does not contribute in hindering the adsorption of ions and to form complexes. The relation between pre-exponential factor and concentration of etchant is very complex and further work is in progress.

Conclusion

(1) Gross dissolution of rhombohedral calcite crystal in HNO_3 acid is maximum at 6 M concentration.
(2) Activation energy and pre-exponential factor increases as concentration of HNO_3 increases.
(3) The gross dissolution of calcite crystal in nitric acid is diffusion controlled.

References

HARTMAN, E., BERGEGI, E., LABAR, J.: J. Cryst. Growth **71**, 191 (1985)
KONNEKE, D., LACMANN, R.: J. Cryst. Growth **46**, 15 (1979)
MENTA, B. J.: Cryst. Res. Technol. **18**, 47 (1983)
MEHTA, B. J., PANDYA, J. R.: Surf. Technol. **15**, 141 (1982)
SIRNOV, E., URUSOVSKAYA, A. J.: Mat. Sci. **15**, 1183 (1980)

(Received March 11, 1987)

Authors' address:

Dr. B. J. MEHTA, G. R. SIDPARA
Solid State Physics and Materials Science Laboratory, Department of Physics
Saurashtra University
Rajkot 360005, India

Book Review

Le Système International d'Unités (SI)

5e Edition

Bureau International des Poids et Mesures. Sèvres. Imprimerie Durand, Luisant; France 1985. 110 p. ISBN 92-822-2092-3

This document is regularly published to inform about resolutions and recommendations of the General Conference on Weights and Measures (CGPM) and of the International Committee for Weights and Measures (CIPM) on the International System of Units. The present edition is a revision of the 4th edition which appeared in 1981. It takes into consideration the decisions of the 17th CGPM and of the CIPM and the amendments made by the Consultative Committee for Units after 1981.

The revision mentioned above covers the definition of the metre, its realization, an explanation of the earlier definition of the sievert and recommendations of the practical realization of the definitions of length, mass, time, electrical quantities, temperature, amount of substance and photometric quantities.

The new decisions are well-founded and a consequence of technological progress. Since basic limits of accuracy have not yet been reached, further revisions are to be expected. In 1985 forty-seven states were members of this convention. In order to make the contents of this editions more accessible for English-speaking readers, an English-Language translation has been included. Due to the difficulty arising from linguistic variations that occur in English-speaking countries this translation, however, is not to be considered as an offical text, whereas the French version is authoritative. Both are part of the present edition. Moreover, translations of this brochure have been published in various other languages. The booklet is well produced and will be indispensable for metrological studies as well as for teaching purposes at universities and technical colleges. It should be available in all libraries collecting literature in science and technology. P. PAUFLER

Cryst. Res. Technol.	22	1987	10	1277–1284

A. D. VLASOV

All-Union Scientific Research Centre for Study of Surface and Vacuum Properties,
State Committee for Standards of the USSR, Moscow

On the Surface Mechanism of Crystal Melting

A crystal melting mechanism is proposed. The melting is caused by enhanced thermal expansion of the lattice on the surface and around structure defects. For the microcrystallites formed this way an approximate calculatian of parameters is made. The microcrystallites surface area per mole is shown to be thousands of m^2, the radius within 1.5 to 10 nm, the gaps between microcrystallites — by 30 percent more than the lattice interatomic distance. The mean thermal linear expansion coefficients determined on the crystal surface are roughly ten times as much as the bulk values.

Предложен механизм плавления кристаллов, в основу которого положено интенсивное тепловое расширение решетки на поверхности и вокруг дефектов структуры. Проведен приближеный расчет параметров микрокристаллитов, образующихся в соответствии с предложенным механизмом. Показано, что молярная поверхность образующихся микрокристаллитов составляет тысячи $м^2$, радиус находится в пределах 1,5—10 нм, расстояние между микрокристаллитами примерно на 30% больше межатомного расстояния. Определенные средние коэффициенты теплового линейного расширения решетки на поверхности кристаллов примерно в 10 раз больше объемных значений.

1. Introduction

The crystal structure is known to be retained, to a certain extent, after crystal melting (see for example ROMANOVA). There is, however, a variety of the liquid structure models (FRENKEL; BAUM). The liquid is thought to be an aggregate of very small crystals with (microcrystallites), or without (cybotaxises, STEWART, MORROW) sharp boundaries, or a very dense gas in which at any instant the domains having a short-range order atomic arrangement (TEMPERLEY) may be present.

The microcrystallite model of liquid allows the crystal and melts properties similarity (ordered atomic arrangements, close values of heat capacity, molar volume etc.). But this model does not completely clear up ways and reason for single-crystal breaking up to very small crystals by melting. That is one of the reasons why this model does not become generally accepted. Besides, there are no data, except the size, on microcrystallites in liquid. The surface energy and area per mole of microcrystallites, their number and gaps betweeen them have not been determined, The present paper is concerned with some of these unsolved problems.

2. The surface melting mechanism

Two conceptions may be regarded as a basis of the melting mechanism suggested. The first one, being the most common, assumes a critical atomic distance at which lengths and angles of atomic bonds in the lattice become uncertain. Hence the lattice becomes unstable and an ordered atomic arrangement is disturbed. The second conception being less known pays attention to the fact that during the heating this critical atomic distance is not reached by all atoms at once, but at first by the atoms located on the inner surfaces (at the lattice defects), because of an intensive surface thermal expansion. Hence at the melting point the area of the inner surfaces increases until the whole crystal is divided into microcrystallites having definite and similar size.

The anomalously intensive thermal expansion of the lattice surface discovered in the last ten years (the coefficients α_s for surface thermal expansion are 2—8 times over the bulk values α_v) (NESTERENKO et al.; FINZEL et al.; FRANK et al.; KIRSCHNER, FEDER) for the elements Ge, Si, W, and compounds GaAs, NaF, and LiF may be attributed to an excitation of Rayleigh's type surface waves. The reason for the excitation of the surface waves, the characteristic frequences of which vary over the range from 10^{11} to 10^{13} Hz, may be thermal vibrations of atoms near the crystal surface (KRASILNIKOV, KRYLOV). Further, it should be noted that excited surface phonons may have relaxation times much exceeding those for phonons in the bulk of the crystal, because scattering of Rayleigh's type surface waves are known to be very slow. If it is true, a thermal energy flow to the surface must exist. That is a very important circumstance! Surface phonons excitation by vibrations of the subsurface atoms may be interpreted as a heat flow directed towards the crystal surface. The heat absorbed by the surface must raise the surface temperature (as compared to the bulk temperature of the crystal) by the value ΔT. On the other hand, a process of surface wave scattering must exist at the same time. This process can be understood as a manifestation of the second law of thermodynamics (the temperature equalization tendency) Hence there are two oppositely directed heat flows: the thermal energy inflow to the surface and the thermal energy outflow from the surface. The rates of these oppositely directed flows obviously depend on the ΔT. Namely, the first flow rate beginning with some value decreases gradually. On the contrary, the second flow rate beginning with zero at $\Delta T = 0$ increases gradually as ΔT increases. So it is clear that at a difference $\Delta T_{st} > 0$ the stationary state at which the rates of two opposite flows are equal to one another will be reached. This stationary state can remain unchanged as long as you like under invariable conditions. The value $\Delta T_{st} = T(n - 1)$ where n is a scale factor which has been calculated (VLASOV 1972, 1986, 1987), T is the absolute temperature in the bulk of the crystal. The value of n is a great one and varies over the range from 1.4 to 8 (VLASOV 1987). Therefore the surface superheating ΔT must be significantly high. But it is not surprising because of a relatively small scattering rate of Rayleigh's waves. The thermal expansion of the lattice surface caused by the intensive thermal vibrations (by surface phonons) must be also several times greater than bulk values, if the α_s are calculated using the crystal bulk temperature.

The anomalously intensive surface expansion caused by the crystal heating is possible until the distance a_{12} between the first and second surface monoatomic layers of the crystal becomes less than the value at which a shear stress resistance vanishes and in accordance with it an excitation of Rayleigh's waves terminates. At this critical distance a **change** of the atomic interaction type which may be caused by a step in the interaction potential curve where $d\varphi/da \approx 0$ (this may be a reason for "dangling" bonds) takes place. The step on the curve may be related to a change of the interaction type or to cooperative phenomena. The assumption of a step-wise curve will not seem strange, if one takes into consideration the fact that a smooth shape of the curve is just a more accustomed than other shapes. Similar assumptions have been discussed earlier (STEWART, MORROW; RASCHEVSKY).

Thus, a cracking action of lattice surface expansion should result in a system of microcracks and eventually, as previously mentioned, leads to breaking the crystal

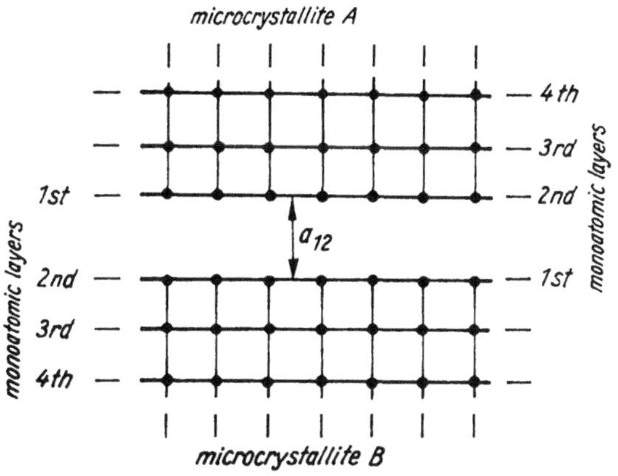

Fig. 1. Mutual arrangement chart for monoatomic layers of neighbouring microcrystallites

bulk into very small crystals. In this process the second monoatomic layers of neighbouring microcrystallites play a role of the first ones for each other (see chart in Fig. 1). As seen in Figure 1, the distance a_{12} between the first and second monoatomic layers is, at the same time, a distance between neighbouring microcrystallites.

When heated at melting point T_m, the crystal is breaking up until the microcrystallite radius reaches a limiting value R. If the microcrystallite radius becomes less than E, the difference between α_s and α_v vanishes, and the distance a_{12} becomes less than a critical one. Then neighbouring microcrystallites unite in a large one. By this process microcrystallite sizes may be regulated and the microcrystallite structure is preserved. A consolidation of microcrystallites to more large species must take place on cooling the melts below T_m as well. Simultaneously with the process of approaching microcrystallite one another a mutual orientation and formation of a continuous crystal lattice must occur. The orientation process is likely to be in the nature of activation. Hence it is clear why a probability of melt supercooling exists and why crystallization takes place when the melt is agitated or shaked.

Finally, it should be noted that the proposed surface mechanism of melting fails for thoroughly perfect crystals. But these crystals are not known to exist in the nature.

3. A method of calculation of the microcrystallite model parameters

These calculation methods are developed insufficiently. The size estimations by X-ray study of liquids (ROMANOVA) are only to be known. The method proposed in present paper is based on a molar surface energy. Two methods of claculation of the surface energy (SKAPSKI; ORIANI, see also PARTINGTON) are known. These methods in principle are correct but give essentialy different results. The first method uses the molar energy U of the crystal lattice, so that

$$\sigma_1 = [(K_v - K_s)/K_v]\, U\,, \tag{1}$$

where K_v, K_s are respectively the coordination numbers for atoms in the bulk and on the surface of the lattice. The first method assumes a bond energies equality for atoms in the bulk and ion the surface of liquids. The second method uses the surface energy σ of liquids:

$$\tilde\sigma_2 = f \cdot \tilde N_A^{1/3} \cdot \tilde v^{2/3} \cdot \sigma = A \cdot \sigma\,, \tag{2}$$

where f is the coefficient depending on a supposed surface layer structure, $\tilde N_A$ is the Avogadro number, $\tilde v$ is the molar volume of liquid. The difference of $\tilde\sigma_1 - \tilde\sigma_2$ is usually positive.

The microcrystallite model of liquids enables us to understand the difference beween $\tilde{\sigma}_1$ and $\tilde{\sigma}_2$ as follows: $\tilde{\sigma}_1$ is the total molar surface energy and $\tilde{\sigma}_2$ is only the part of $\tilde{\sigma}_1$ required to transfer an atom from the microcrystallite surface to the surface of liquid (see Fig. 2). It will be easily understood that $\tilde{\sigma}_2$, being determined from experimental values of σ by equation (2), are energies needed to take on the surface of liquid a part of the microcrystallite surfaces biut not isolated atoms. So molar surface energy $\tilde{\sigma}_{\mathrm{mc}}$ of microcrystallites equals to

$$\tilde{\sigma}_{\mathrm{mc}} = \tilde{\sigma}_1 - \tilde{\sigma}_2 \,. \tag{3}$$

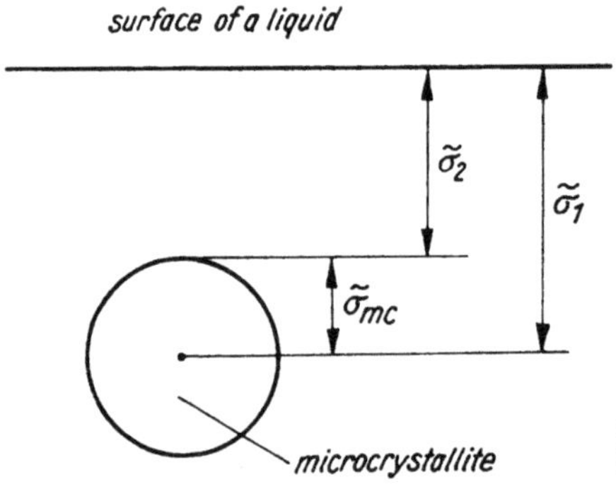

Fig. 2. Three distinguished surface energies ($\tilde{\sigma}_1$, $\tilde{\sigma}_2$, $\tilde{\sigma}_{\mathrm{mc}}$) for the microcrystallite model of liquid

It should be noted that $\tilde{\sigma}_1$ and $\tilde{\sigma}_2$ are free surface energies. But if one adds entropy contributions to $\tilde{\sigma}_1$ and $\tilde{\sigma}_2$, the difference $\tilde{\sigma}_1 - \tilde{\sigma}_2$ will be changed by ± 10 percent. Due to estimation uncertainties we decided not to account for these contributions at all.

Further

$$\sigma_{\mathrm{mc}} = \tilde{\sigma}_{\mathrm{mc}}/A_{\mathrm{mc}} \,, \tag{4}$$

where $A_{\mathrm{mc}} \approx A$ (see equation (2)). Now from molar heat melting $\tilde{q}_{\mathrm{m}}$ a microcrystallite surface area per one mole of microcrystallites may be determined

$$\tilde{S} = \tilde{q}_{\mathrm{m}}/\sigma_{\mathrm{mc}} \tag{5}$$

if the total of $\tilde{q}_{\mathrm{m}}$ is consumed for microcrystallite surface formation.

On the other hand, the surface area $\tilde{S} = \tilde{n} \cdot 4\pi R^2$, where $\tilde{n}$ is microcrystallite number per mole of liquid, R is their radius. Further, the molar volume $\tilde{v}$ of liquid $\tilde{v} = \tilde{n}(4/3)\,\pi R^3$. Hence, the number n and radius R may be estimated. Now, knowing $\tilde{S}$ and supposing that the volume increase $\Delta\tilde{v}_{\mathrm{m}}$ on melting is only due to an increase of the intermicrocrystallite gaps, as compared to the lattice parameter a, we find:

$$\Delta a = \Delta\tilde{v}_{\mathrm{m}}/(1/2)\,\tilde{S} \,, \tag{6}$$

where the coefficient (1/2) is included in order not to count twice the contacting microcrystallite surfaces.

Initial data for estimation of different microcrystallite parameters are given in Table 1, where the values are presented for metals having supposed bcc- or fcc-structure (short-range order) in the liquid state and also for Hg. The molar volume $\tilde{v}$ and lattice energies $\tilde{U}$ for solid metals have been extrapolated to the melting points.

Table 1
Data for evaluation of microcrystallite parameters

Lattice type, f, $(K_v - K_s)/K_v$	element	a, 0.1 nm (Smithells)	$\tilde{v}$/cm³/mole (Kikoin)	$\sigma/10^{-7}$ J/cm² (Nizhenko, Floka)	$\tilde{U}$, kJ/mole (Kikoin)	$\tilde{q}_m$, kJ/mole (Smithells)	$\Delta\tilde{v}_m$, cm³/mole (Smithells, Kikoin)	α_v, 10^{-5} degree^{-1} (Smithells, Wilson, Touloukian)
	Li	3.03	13.0	470	160	2.89	0.21	5.6
	Na	3.71	24.2	194	101	2.64	0.60	7.6
	K	4.62	46.1	107	89.5	2.34	1.18	8.3
BCC	Rb	4.81	58.3	85.7	76	2.20	1.46	9.0
1.12	Cs	5.24	69.9	68.8	78.5	2.10	1.82	9.7
0.36	Fe	2.48	7.83	1800	367	15.2	0.27	2.3
(Oriani)	Tl	3.36	17.7	460	176	4.30	0.39	3.7
	Al	2.86	10.6	860	315	10.5	0.69	3.7
	Ni	2.49	7.11	1800	400	17.7	0.32	2.2
FCC	Cu	2.55	7.62	1300	327	13.0	0.32	2.6
1.09	Ag	2.88	11.1	926	262	11.4	0.42	2.3
0.25	In	3.30	16.0	555	241	3.27	0.32	4.4
(Oriani)	Pt	2.77	9.7	1746	531	19.7	0.56	1.6
	Au	2.88	10.8	1130	364	12.8	0.55	2.3
	Pb	3.49	18.8	445	193	4.98	0.66	3.7
Rhombohedral 1.04 0.5 (Oriani)	Hg	3.00	14.1	485	61.1	2.32	0.52	6.1

Table 2
Evaluated microcrystallite parameters

Lattice type	element	$\tilde{\sigma}_2$, kJ/mole	$\tilde{\sigma}_{mc}$, kJ/mole	σ_{mc}, 10^{-7} J/cm^2	$\tilde{S}$, 10^7 cm^2/mole	R, 0.1 nm	RT_m, 10^{-4} cm · degree	$\tilde{n}$, 10^{20} mole^{-1}	N	Δa, 0.1 nm	$\frac{a+\Delta a}{a}$	α_s/α_v
BCC	Li	24.5	33.0	632	4.57	85	3.8	0.15	4500	0.92	1.30	12
	Na	15.3	21.1	267	9.89	73	2.7	0.45	1600	1.2	1.33	12
	K	13.1	19.1	156	15.0	92	3.1	0.42	1600	1.6	1.34	12
	Rb	12.2	15.2	107	20.6	85	2.6	0.68	1000	1.4	1.29	10
	Cs	11.0	17.3	108	19.4	108	3.3	0.40	1700	1.9	1.36	12
	Fe	67.1	65.2	1750	8.70	27	4.9	2.8	230	0.62	1.24	5.8
	Tl	29.5	33.9	528	8.14	65	3.8	0.46	1400	0.92	1.27	13
FCC	Al	38.2	40.5	912	11.5	28	2.6	3.6	120	1.2	1.42	12
	Ni	61.2	39.0	1150	15.4	14	2.4	19	23	0.40	1.16	4.2
	Cu	46.3	35.5	997	13.0	18	2.3	10	440	0.50	1.20	5.7
	Ag	42.4	23.1	504	22.6	15	1.8	25	17	0.38	1.13	3.6
	In	32.4	28.0	479	6.83	70	3.0	0.33	130	0.94	1.32	17
	Pt	73.0	60.0	1440	13.7	21	4.3	7.3	60	0.82	1.30	9
	Au	50.8	40.2	893	14.3	23	3.0	6.7	64	0.76	1.25	8.4
	Pb	28.9	19.4	298	16.7	34	2.0	3.5	120	0.80	1.23	10
rhombohedral	Hg	24.9	5.65	110	21.1	20	0.47	12	60	0.49	1.16	11

4. Results and discussion

The results of calculation for the microcrystallite parameters are given in Table 2. As seen in Table 2, the molar surface energies $\tilde{\sigma}_2$ of liquid metals and $\tilde{\sigma}_{mc}$ of microcrystallites are roughly equal. The surface area $\tilde{S}$ of microcrystallites is enormous and makes up thousands of m^2/mole. According to X-ray data for liquid metals (ROMANOVA) the minimum microcrystallite radius amounts to ~2.5 nm for Pb, ~1.8—2.0 nm for Au, Cu, Tl, In, ~ 1.5 nm for Fe. These data are in agreement or lower than the data presented respectively in Table 2: 3.4, 2.3, 1.8, 6.5, 7.0, 2.7 nm. The radius R diminishes as the melting point T_m increases so that the product RT_m is roughly constant and is equal to $3 \cdot 10^{-4}$ cm · degree. The microcrystallite number $\tilde{n}$ changes considerably versus R. The elemental cells number N in a microcrystallite estimated by the ratio $\tilde{N}_A/\tilde{n}$ differ very much for individual metals. The reason for that is not clear.

An important result derived from the calculation lies in that the ratio $\Delta a/a \approx 0.28$ and is approximately the same for different metals. The distance $a + \Delta a$ is mentioned above a_{12} (see (Fig. 1). At the distance equal to or greater than $a + \Delta a$ the bond angles become uncertain, which results in a free microcrystallite migration and in small values of melts viscosity with preserved unoriented bonds between the microcrystallites. Therefore, liquids are characterized by a tensile strength. For example, water and some of organic liquids have the tensile strength of $(1.5-3.0) \cdot 10^7$ Pa (FLÜGGE). As seen from the last column of Table 2, the ratio $\alpha_s/\alpha_v = \Delta a/aT_m\alpha_v$ is within 3.6—17, the mean ratio being 9.8. These values are consistent with the direct determination results: $\alpha_s/\alpha_v = 2-8$ (references are mentioned above).

Besides of the metals that do not change the structure on melting, a similar calculation was made for Ga, P, S, Se, Te. The results are like the presented in Table 2, and R is equal respectively to 7.0, 5.7, 9.4, 2.6, and 1.3 nm.

References

BAUM, B. A.: Metallicheskie zhidkosti, Moscow 1979 (in Russian)

FINZEL, H.-U., FRANK, H., HOINKES, H., LUSCHKA, M., NAHR, H., WILSCH, H., WONKA, U.: Surf. Sci. **49**, 577 (1975)

FLÜGGE, S., ed.: Handbuch der Physik. B. 6, Berlin 1958, p. 551

FRANK, H., HOINKES, H., WILSCH, H.: Surf. Sci. **64**, 362 (1977)

FRENKEL, Ya. I.: Kineticheskaya teoriya zhidkostei, Leningrad 1975 (in Russian)

KIKOIN, I. K., ed.: Tablitsy fisicheskikh velichin, Moscow 1976 (in Russian)

KIRSCHNER, J., FEDER, R.: Surf. Sci. **104**, 448 (1981)

KRASILNIKOV, V. A., KRYLOV, V. V.: Vvedenie v fizicheskuyu akustiku, Moscow 1984 (in Russian)

NESTERENKO, B. A., BORODKIN, A. D., SNITKO, O. V.: FTT **15**, 2602 (1973) (in Russian)

NIZHENKO, V. I., FLOKA, L. I.: Poverkhnostnoye natyazheniye zhidkikh metallov i splavov, Moscow 1981 (in (Russian)

ORIANI, R. A.: J. Chem. Phys. **18**, 575 (1950)

PARTINGTON, J. R.: An advanced treatise on physical chemistry. Vol. 2, London 1955

RASCHEVSKY, N. V.: Z. Phys. **40**, 214 (1927)

SKAPSKI, A. S.: J. Chem. Phys. **16**, 386 (1948)

SMITHELLS, C. I., ed.: Metals Reference Book. London & Boston 1976

STEWART, G. W., MORROW, R. M.: Phys. Rev. **30**, 232 (1927)

TEMPERLEY, H. N. V.: Changes of States, London 1956

TOULOUKIAN, Y. S., ed.: Thermophysical properties of matter, Vol. 12, New York 1975

Vlasov, A. D.: Zhurn. Fiz. Khim. **46**, 1661 (1972) (in Russian)
Vlasov, A. D.: Dokl. AN SSSR **287**, 892 (1986) (in Russian)
Vlasov, A. D.: Cryst. Res. Technol. **22**, 527 (1987)
Wilson, D. R.: Struktura zhidkikh metallov i splavov, Moscow 1972 (in Russian)

(Received April 4, 1987)

Author's address:

Dr. A. D. Vlasov
All-Union Scientific Research Centre for Study of Surface and Vacuum Properties
State Committee for Standards of the USSR
Leninski prospekt 9
117049 Moscow
USSR

Cryst. Res. Technol.	22	1987	10	1285–1288

K. J. PRATAP, M. NAGABHUSHANAM, A. BALARAMAIAH, V. HARI BABU

Department of Physics, University College of Science
Osmania University, Hyderabad, India

Electrical Transport Properties of CdTe:Sb Films

Hall coefficient and dc conductivity measurements are made on p-type CdTe:Sb films grown by vacuum evaporation technique on glass substrate. The grain boundary potential barrier, which is found mainly to limit the mobility of carriers is calculated as a function of film thickness. The n-type conduction is found to dominate over p-type conduction above about 330 K. The ratio of electron to hole mobility is also calculated.

Introduction

CdTe thin films are of general interest because of their potential application in thin-film solar cells (TYAN et al.; BARBEE et al.). These applications in turn depend on various scattering mechanisms limiting the carrier mobility, and the conditions under which the films are prepared (WIEDER, CALASON). For epitaxial films it is the dislocation scattering (WIEDER), which is more important whereas for polycrystalline films it is the grain boundary potential barrier scattering (WAXMAN et al.) which is more relevant for the transport properties. Surface scattering (RICHARDS) mainly limits the mobility of free carriers in films having dimensions comparable to the carrier mean free path. The basic interest in the field of electrical transport properties of thin films is mainly due to the fact that this will lead to a better understanding of the surface properties of bulk or thin film semiconductor devices.

Different workers (WAXMAN et al.; VAN HEEK) have found an exponential temperature dependence of grain-boundary barrier-limited mobility in polycrystalline CdS and CdSe films respectively. However, these are wide-gap semiconductor materials having nearly parabolic band structure. The validity of the existence of a potential barrier as proposed by PETRITZ is established only for nondegenerate semiconductors having wide energy gap and parabolic band structure. With an interest to examine whether this applies to p-type CdTe:Sb, the present investigations are taken up. This paper reports measurements on the Hall coefficient and dc conductivity on p-type CdTe:Sb films for different thickness (1020, 1840, and 2860 Å).

Experimental

CdTe:Sb thin films were prepared by sequential vacuum deposition of CdTe and Sb on clean and hot glass substrates held at 210 °C at a constant rate of deposition in a vacuum of 5×10^{-5} torr. To achieve uniform doping the films were annealed at high temperature for 15 minutes. The films were subsequently aged for more than 48 hours in a dessicator to anneal out residual defects. Films of different thickness were prepared by controlling the evaporation time.

For ohmic contacts, high purity Al was evaporated on the films under high vacuum. Hall coefficient and conductivity were measured using five probe technique. All the measurements were carried in vacuum from 77—350 K with help of a specially made cryostat, the description of which was given else where (Nagabhushanam). The current and the magnetic field ($B = 7$ KG) directions were reversed to eliminate the errors due to thermomagnetic effects. The Hall coefficient was found to be independent on the magnetic field and the sample current. The Ohmic nature of the contacts was confirmed by the I—V characteristics throughout the temperature range.

Results and discussion

The variation of Hall coefficient R_H with temperature T (log R_H vs $1/T$) for all the three films of thickness 1020, 1840, 2860 Å are shown in Figure 1. These samples are referred as I, II, and III, respectively. It is observed that in the low temperature region the Hall coefficient decreases slowly with the increase of temperature, which is typical for a nearly degenerate extrinsic semiconductor. At temperatures above about 170 K for all the films the Hall coefficient falls rapidly (and approaches zero, which is also seen in Figure 1). This shows that at high temperatures besides the holes the free electrons are also contributing to the conduction process. The value of R_H is given by

$$R_H = r_H \frac{1 - xb^2}{\exp (1 + xb)^2}, \tag{1}$$

where b is the ratio of the electron to hole mobility (μ_n/μ_p), x is the ratio of the concentration of electrons (n) to that of holes (p) and $r_H = \langle\tau^2\rangle/\langle\tau\rangle^2$ is the Hall factor which is approximately equal to 1 (Naidu, 79). The Hall coefficient rises again as the temperature increases further showing the predominance of the free-electron conduction process, which is due to the high mobility of electrons as compared to that of holes, inspite of the fact that $n < p$. At the temperature where R_H goes through zero, the contributions of electrons and holes to the conduction process cancel out and $p = nb$.

The temperature variation of R_H above 330 K corresponds to the intrinsic behaviour. It can also be seen from Figure 1 that the inversion temperature (where $R_H \to 0$)

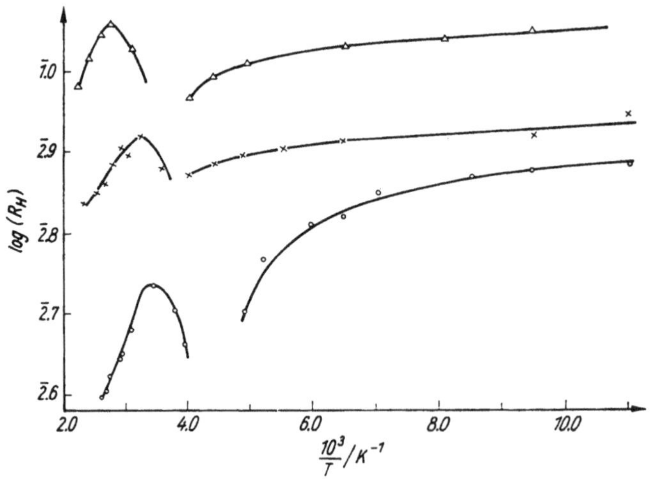

Fig. 1. Variation of Hall coefficient with temperature of CdTe:Sb films of different thickness: ○ 1020 Å, × 1840 Å, △ 2860 Å

increases with the film thickness. This is because of the fact that as the film thickness increases, there must be a large number of free holes to be compensated by free electrons to reach the inversion stage. The value of b can be calculated by the expression:

$$\frac{R_{max}}{R_{ex}} = \frac{(b-1)^2}{4b}, \tag{2}$$

where R_{max} is the maximum value of the Hall coefficient (algrebraically) and R_{ex} is the extrinsic value of the Hall coefficient. The values of b obtained for various films are shown in Table 1.

Table 1

Thickness of the film	inversion temperature	b	Ø mev
1020 Å	220	4.6	4.0
1840 Å	260	5.7	2.6
2860 Å	285	7.9	1.3

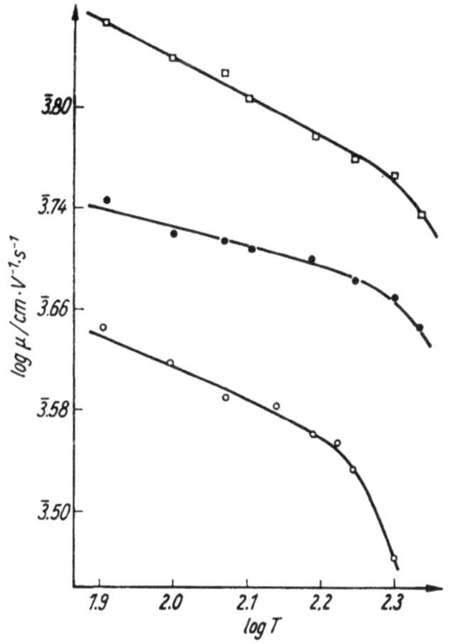

Fig. 2. Variation of Hall mobility with temperature of CdTe:Sb films of different thickness: ○ 1020 Å, ● 1840 Å, ● 2860 Å

The variation of Hall mobility μ_H ($\mu_H = R_H\sigma$) with temperature is shown in Figure 2. The mobility for all the three films is found to decrease with the increase of temperature. This shows that the lattice mobility μ_l becomes more important as compared to the grain boundary limited mobility μ_B. The value of μ_B has been calculated from the observed Hall mobility data in the following manner.

$$\frac{1}{\mu_H} = \frac{1}{\mu_d} + \frac{1}{\mu_B}, \tag{3}$$

where μ_d is the diffused scattering mobility given by SONDHEDMER

$$\mu_d = \frac{\mu_1}{1 + \frac{3}{8}r}. \tag{4}$$

r being the ratio of the film thickness to the mean free path.

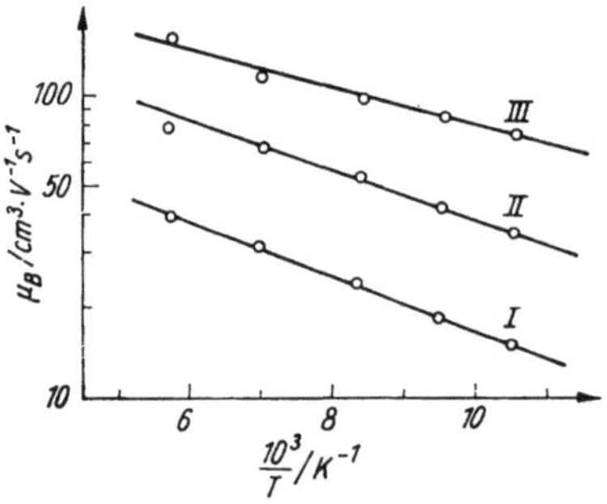

Fig. 3. Variation of grain boundary mobility with temperature of CdTe:Sb films of different thickness

Temperature variation of μ_B ($\log \mu_B$ vs $10^3/T$) calculated from equation (3) is shown in Figure 3. The mobility μ_B is found to increase with temperature exponentially for all the three films. The dependence of μ_B on the grain barrier potential is expressed as (TYAGI, 1975).

$$\mu_B = \frac{q v_{th}}{4NRT} \exp\left(-\frac{e\Phi}{RT}\right) \quad (5)$$

where N is the number of crystallites per unit length and V_{th} is the average thermal velocity. The values of the barrier potential Φ for the three films calculated using equation (5) are shown in Table 1. It is seen from the table that the value of Φ increases as the thickness decreases. The low value of barrier potential indicates that small notch-like barriers are present in the films investigated. Similar type of barriers were also observed in InSb by SHIGITA et al.).

We thus conclude that in p-type CdTe:Sb films a single type of potential barrier associated with grain boundary dominates the conduction and the barrier leads to a mobility which varies exponentially with temperature.

The authors are thankful to the Head, Department of Physics, Osmania University, for his constant encouragement and authors Pratap and Balaramaiah are thankful to the U.G.C., New Delhi for providing F.I.P.

References

BARBEE, M., BAILLY, F., LUXOT, D., COHEN-SOLAL, G.: 16th IEEE Photovoltaic Specialists Conference Proceedings, IEEE, New York 1982, p. 1133

NAGABHUSHANAM, M.: Thesis, submitted to Osmania University 1982, p. 57

PETRITZ, R. L.: Phys. Rev. **79**, 1023 (1950)

RICHARDS, J. L.: The use of thin films in physical investigations, London 1966, p. 419

SEGALL, B., LORENZ, M. R., HALSTED, R. E.: Phys. Rev. **129**, 2471 (1963)

SHIGITA, J., KOTERA, N., OI, T.: J. Appl. Physics **47**, 621 (1976)

SONDHEDMER, E. H.: Adv. Phys. **1**, 1 (1952)

TYAGI, M. S.: phys. stat. sol. (a) **30**, 609 (1975)

TYAN, Y. S., PEREZ-ALBUERNE, E. A.: 16th IEEE photovolataic Specialists Conference Proceedings, IEEE, New York 1982, p. 794

VAN HEEK, H. F.: Solid State Electronics **11**, 459 (1968)

WAXMAN, A., HENRICH, V. E., SHALKROSS, F. V., BORKAN, H., WEIMER, P. K.: J. Appl. Physics **36**, 168 (1965)

WIEDER, H. H.: Solid State Electronics **9**, 373 (1966)

WIEDER, H. H., CALASON, A. P.: Solid State Electronics 8, 467 (1965)

(Received February 24, 1987)

Authors' address:

Professor V. HARI BABU, K. J. PRATAP, M. NAGABHUSHANAM, A. BALARAMAIAH
Department of Physics, University College of Science,
Osmania University
Hyderabad-500007, India

Cryst. Res. Technol.	22	1987	10	1289–1296

R. Assenov, V. A. Moshnikov, B. Patarov, D. A. Yaskov

Semiconductor Physics and Technology Research Laboratory, University of Sofia, Bulgaria, and Leningrad Electrotechnical Institute, USSR

Method of X-Ray Spectral Microanalysis of PbTe and SnTe Obtained by the Iodide Method

This an examination of the characteristics of the X-ray spectral microanalysis of the tellurides of lead and thin, obtained by the iodide method. It was demonstrated that applying the method of calculating the composition in relation to the relative intensity of the characteristic X-ray radiations of the analytical lines TeL_α and IL_α secures the possibility of analyzing relief surfaces. Besides that, the method makes it possible to make calculations without using numerous traditional semi-empirical expressions under conditions of changes in the values of the accelerating voltage.

Рассмотрены особенности количественного рентгеноспектрального микроанализа теллуридов свинца и олова, полученных иодидным методом. Показано, что применение методики расчета состава по отношению относительных интенсивностей характеристических рентгеновских излучений аналитических линий TeL_α и IL_α обеспечивает возможность анализа рельефных поверхностей. Кроме того разработанные методики позволяют проводить расчет без использования многих традиционных полуэмпирических выражений в условиях изменения значения ускоряющего напряжения.

It has been demonstrated in publications by Moldovanova et al. that the iodide method of synthesizing samples of lead telluride and tin telluride with outstanding electrophysical properties is a promising one.

Gafni et al. carried out experiments on the growth of solid solutions of $Pb_{1-x}Sn_xTe$ by the iodide technology. The essence of this technology was presented by Kwestroo, and it consists in carrying out a process of interaction between Pb(Sn) and Te in a solvent which is iodide of lead or tin. This leads to essential decrease in the temperature of the synthesis and to avoiding the typical pollutions of the material as a result of the sufficiently high reaction temperatures. The respective tellurides are obtained in powder form after the separation of the iodides in a flow of inert gas or under the conditions of dynamic vacuum. The quality of the material was controlled by electrophysical methods. The nature of the changes in the electrophysical properties has been examined by Assenov et al.

It becomes clear from what has been stated above that it is necessary to control the degree of separation of the iodides from the reaction products. As already indicated, some of the effective methods could be X-ray phase analysis (Assenov et al.) and the X-ray spectral microanalysis (Assenov et al.). The application of the X-ray spectral analysis becomes increasingly promising at the stage of working out the technology of obtaining thin layers or of the growing of monocrystals, because this method makes it possible to determine the composition in local areas and to assess the distribution of the composition in the most important technological directions.

At the same time, the traditional methods of conducting X-ray spectral microanalysis, which are based on recording the effects of generation and absorption of the characteristic X-ray radiation, and also the fluorescence from the characteristic and uninterrupted spectral method (the ZAF method), operate satisfactorily only for samples of smooth surface (REED). The application of the ZAF method is difficult under conditions when powdered material is analyzed.

The aim of the work undertaken was to examine the possibility of using the method of relative intensities (MOSHNIKOV; DEDEGBAEV et al.) in analyzing the tellurides obtained by the iodide method. The method of relative intensities may be used in the analysis of solid solutions containing elements that are close in their atomic numbers, when there is no fluorescence in the system.

The semiconductor solid solutions exist in quasibinary sections. By way of example, let us take $A_{1-x}C_xB$. The formula given signifies that the x composition is univalently connected with the concentrations of the components. Consequently, the ratio of the concentrations of two arbitrarily taken elements is likewise univalently connected with the composition $C_i/C_j = f(x)$. This dependence makes it possible to expand the possibilities of X-ray spectral microscopic analysis of the composition by the ratio of the relative intensities (DEDEGKAEV et al.) in the abscence of fluorescence effect. In this case the ratio of the relative intensities tends toward C_i/C_j. This means that it is practically independent on the values of the accelerating voltage U_0 and of the angle of selection of the X-ray radiation θ. Besides that, the ratio of the relative intensities, as indicated by DEDEDKAEV et al., is independent on the thickness of the sample investigated. This makes it possible to use it in the analysis of both voluminal and of thin-layer objects by a unified method.

The small dependence of the method of relative intensities on the sample thickness and on the selection angle of the X-ray radiation (i.e. on the relief of the surface) makes this method a promising one in the analysis of materials obtained by the iodide method. What remains to be done is to clarify the problem on the applicability of the method of relative intensities for analyzing the Sn–Te–I and Pb–Te–I systems containing elements which are close in their atomic number (Sn, Te, I).

Since in the case of applicability of the method of relative intensities Φ_j^i is approximately equal to C_i/C_j, its applicability may be evaluated by means of calculation. Assuming that the object is of the flat-surface type, the analysis may be performed by the traditional ZAF method:

$$\frac{I_i^{sp}}{I_i^{st}} = \frac{C_i^{sp}}{C_i^{st}} P_Z P_A P_F P_T ,$$

where I_i^{st}, I_i^{sp} is the intensity of the characteristic X-ray radiation of the analytical line of the i-th element in the standard with known content and, accordingly, in the sample: C_i^{st}, C_i^{sp} is the mass concentration of the i-th element in the standard and in the sample, accordingly; P_Z, $P_A P_F P_T$ are the correction coefficients taking account of the effects of excitation (P_Z) and of absorption (P_A) of the characteristic X-ray radiation of the i-th element, as well as the fluorescence caused by the characteristic (P_F) and the continuous (P_T) spectra.

Then the ratio of the relative intensities of the i-th and j-th elements may be written as follows:

$$\Phi_j^i = \frac{C_i^{sp}}{C_i^{st}} \frac{C_j^{sp}}{C_j^{st}} \frac{P_{Zi} P_{Ai} P_{Fi} P_{Ti}}{P_{Zj} P_{Aj} P_{Fj} P_{Tj}} ,$$

where

$$\Phi_j^i = \left(\frac{I_i^{sp}}{I_i^{st}}\right) \Big/ \left(\frac{I_j^{sp}}{I_j^{st}}\right).$$

It is obvious that

$$\lim_{Z_i \to Z_j} \Phi_j^i = \frac{C_i^{sp}}{C_i^{st}} \frac{C_j^{st}}{C_i^{sp}}.$$

In the cases when there is no fluorescent effect this expression is sufficiently correct for the analytical lines of elements with close atomic numbers Z_i and Z_j.

By way of example, let us examine the system $(1 - x)$ SnTe–xI and let us find the connection between the composition X and Φ_j^i under the assumption that the method of relative intensities is applicable. The mass concentrations of tin (C_{Sn}^{sp}), tellurium C_{Te}^{sp} and iodine in the sample may be written as follows:

$$C_{Sn} = \frac{(1 - x) A_{Sn}}{M(x)}, \qquad C_{Te} = \frac{(1 - x) A_{Te}}{M(x)}, \qquad C_I = \frac{(1 - x) A_I}{M(x)},$$

where

$$M(x) = (1 - x) A_{Sn} + (1 - x) A_{Te} + x A_I .$$

A_i is the atomic mass of the i-th element.

As a rule, binary compounds are used as standards in the X-ray spectral microscopic analysis of semiconductor solid solutions. There is a whole series of binary compounds for the systems of Sn–Te–I and Pb–Te–I. The optimum among them for the determination of φ_i^j is the selection of the components of iodine (j) and tellurium (i), e.g. the compound TeI_4. This is explained by the proximity of their atomic numbers ($Z_I = 53$, $Z_{Te} = 52$), and also by the fact that upon reducing the composition of X the character of the change in the mass concentrations of iodine and of tellurium is of a contrary nature. This secures higher sensitivity in determining the change in the ΔX composition. The standard used in the analysis was TeI_4, but this role may be performed also by the TeI standards (Tab. 1), and also by ternary compounds containing Te and I. The mass concentrations C_I^{st} and C_{Te}^{sp} in the TeI_4 standard are equal to

$$C_I^{st} = \frac{4A_I}{A_{Te} + 4A_I}, \qquad C_{Te}^{st} = \frac{A_{Te}}{A_{Te} + 4A_I}.$$

When the method of relative intensities is applicable, the expression Φ_j^i acquires the following form:

$$\Phi_I^{Te} \approx \frac{4(1 - x)}{x}.$$

Consequently,

$$x \approx \frac{4}{4 + \Phi_i^{Te}}$$

and this ratio must not depend in the angle of selection of the X-ray radiation θ or on the value of the accelerating voltage U_0.

The applicability of the method of relative intensities can be readily examined by comparing the values Φ_j^i from the results for I_i^{sp}/I_i^{st} and I_j^{sp}/I_j^{st}, obtained by the ZAF method at different values of X, U_0 and θ, whereby

$$I_i^{sp}/I_i^{st} = f(U_0, \theta) .$$

Table 1 contains the results for $I_{Te}^{sp}/I_{Te}^{TeI_4}$; I_{Te}^{sp}/I_{Te}^{TeI}; $I_I^{sp}/I_I^{TeI_4}$; I_I^{sp}/I_I^{TeI}; Φ_I^{Te}, and X at a value of the accelerating voltage $U_0 = 15$ kV and $\theta = 18°$ (technical parameter of the microanalyzer M_s-46).

The calculations were made on a CM-1 computer. The corrective functions of the effects of excitation and absorption were taken according to PHILLIBERT, taking account of Lenard's coefficient as modified by HEINRICH. The fluorescence effect was

Table 1
Condition of the analysis: $U = 15$ kV; $B = 18$ deg

Parameters	Composition X, Mol. fraction I										
	0	0.1	0.2	0.3	0.4	0.5	0.6	0.7	0.8	0.9	1.0
I^{Sp}_{Te}/I^{Te}_{Te}	0.49623	0.47042	0.44170	0.40956	0.37333	0.33219	0.28508	0.23057	0.16679	0.09115	0.00000
$I^{Sp}_{Te}/I^{TeI_4}_{Te}$	2.47163	2.34307	2.20002	2.03994	1.85949	1.65457	1.41993	1.14842	0.83075	0.45400	0.00000
I^{Sp}_{I}/I^{I}_{I}	0.00000	0.05192	0.10971	0.17440	0.24733	0.33017	0.42508	0.53492	0.66350	0.81606	1.00000
$I^{Sp}_{I}/I^{TeI_4}_{I}$	0.00000	0.06497	0.13730	0.21826	0.30953	0.41320	0.53198	0.66945	0.83037	1.02130	1.25150
φ^{Te}_{I}		36. 0594	16. 0231	9.34629	6.00740	4.00428	2.66914	1.71546	1.00045	0.44453	0.00000
$4/4 + \Phi^{Te}_{I}$	0.00000	0.09985	0.19976	0.29970	0.39970	0.49973	0.59978	0.69985	0.79992	0.89998	1.00000
I^{Sp}_{Te}/I^{TeI}_{Te}	0.99012	0.93862	0.88132	0.81724	0.74490	0.66281	0.56881	0.46005	0.33279	0.18187	0.00000
I^{Sp}_{I}/I^{TeI}_{I}	0.00000	0.10414	0.22006	0.34982	0.49611	0.66228	0.85266	1.07299	1.33091	1.63693	2.00598
Φ^{Te}_{I}		9.01305	4.00490	2.33617	1.50148	1.00080	0.66710	0.42875	0.25004	0.11110	0.00000
$1/1 + \Phi^{Te}_{I}$	0.00000	0.09986	0.19980	0.29974	0.39976	0.49980	0.59984	0.69991	0.79997	0.90000	1.00000

Table 2
Condition of the analysis

x	$\theta = 18$					$U = 20$ kV		
	$U = 15$ kV	$U = 20$ kV	$U = 25$ kV	$U = 30$ kV	$U = 35$ kV	$\theta = 13$	$\theta = 23$	$\theta = 43$
1	2	3	4	5	6	7	8	9
	System: $(1 - x)$ SnTe–xI					Sample TeI_4 $x = 4/4 + \Phi$		
0.1	0.09985	0.09978	0.09974	0.09970	0.09969	0.09896	0.10031	0.10131
0.2	0.19976	0.19964	0.19956	0.19951	0.19948	0.19820	0.20057	0.20234
0.3	0.29970	0.29957	0.29946	0.29940	0.29936	0.29768	0.30071	0.30309
0.4	0.39970	0.39956	0.39944	0.39968	0.39934	0.39740	0.40094	0.40357
0.5	0.49973	0.49959	0.49950	0.49943	0.49940	0.49736	0.50103	0.50375
0.6	0.59978	0.59937	0.59959	0.59955	0.59951	0.59754	0.60105	0.60363
0.7	0.69985	0.69977	0.69972	0.69969	0.69967	0.69722	0.70097	0.70321
0.8	0.79992	0.79989	0.79986	0.79985	0.79984	0.79848	0.80079	0.80248
0.9	0.89998	0.89997	0.89997	0.89996	0.89997	0.89918	0.90048	0.90141
	System: $(1 - x)$ SnTe–xSnI					Sample TeI_4 $x = 4/4 + \Phi$		
0.1	0.09984	0.09978	0.09973	0.09969	0.09967	0.09896	0.10030	0.10130
0.2	0.19973	0.19961	0.19952	0.19947	0.19943	0.19816	0.20055	0.20232
0.3	0.29967	0.29951	0.29940	0.29931	0.29928	0.29761	0.30072	0.30305
0.4	0.39963	0.39945	0.39931	0.39925	0.39920	0.39728	0.40085	0.40349
0.5	0.49962	0.49944	0.49930	0.49923	0.49919	0.49719	0.50089	0.50364
0.6	0.59964	0.59947	0.59935	0.59928	0.59924	0.59731	0.60088	0.60359
0.7	0.69970	0.69956	0.69945	0.69939	0.69935	0.69767	0.70077	0.70308
0.8	0.79978	0.79966	0.79960	0.79955	0.79952	0.79823	0.80061	0.80239
0.9	0.89987	0.89982	0.89978	0.89975	0.89974	0.89901	0.90034	0.90131
	System: $(1 - x)$ SnTe–XSNI$_2$					Sample TeI_4 $x = 2/2 + \Phi$		
0.1	0.09985	0.09978	0.09974	0.09973	0.09969	0.09898	0.10031	0.10131
0.2	0.19977	0.19985	0.19957	0.19952	0.19949	0.19821	0.20057	0.20235
0.3	0.29971	0.29957	0.29947	0.29941	0.29938	0.29769	0.30079	0.30310
0.4	0.39970	0.39955	0.39945	0.39937	0.39905	0.39740	0.40094	0.40356
0.5	0.49971	0.49956	0.49945	0.49940	0.49937	0.49733	0.50100	0.50372
0.6	0.59974	0.59961	0.59951	0.59946	0.59943	0.59747	0.60099	0.60358
0.7	0.69978	0.69968	0.69961	0.69956	0.69954	0.69880	0.70090	0.70315
0.8	0.79985	0.79977	0.79972	0.79969	0.79967	0.79836	0.80069	0.80240
0.9	0.89971	0.89987	0.89986	0.89983	0.89982	0.89906	0.90039	0.90136
	System: $(1 - x)$ SnTe–xSnI$_4$					Sample TeI_4 $x = 1/1 + \Phi$		
0.1	0.09987	0.09981	0.09976	0.09973	0.09972	0.09831	0.10033	0.10132
0.2	0.19981	0.19971	0.19964	0.19960	0.19957	0.19828	0.20063	0.20239
0.3	0.29978	0.29967	0.29959	0.29954	0.29952	0.29780	0.30087	0.30316
0.4	0.39978	0.39968	0.39959	0.39956	0.39962	0.39755	0.40105	0.40364
0.5	0.49980	0.49971	0.49964	0.49959	0.49957	0.49748	0.50113	0.50381
0.6	0.59983	0.59976	0.59969	0.59966	0.59964	0.59763	0.60111	0.60368
0.7	0.69988	0.69981	0.69976	0.69973	0.69971	0.69795	0.70099	0.70324
0.8	0.79991	0.79986	0.79965	0.79985	0.79983	0.79847	0.80078	0.80246
0.9	0.89995	0.89995	0.89992	0.89992	0.89990	0.89913	0.90044	0.90137

Table 2 (continued)

1	2	3	4	5	6	7	8	9
	System: $(1-x)$ SnTe–xTeI					Sample TeI_4 $x = 4/\Phi$		
0.1	0.09985	0.09976	0.09970	0.09966	0.09965	0.09888	0.10036	0.10147
0.2	0.19972	0.19957	0.19947	0.19942	0.19938	0.19779	0.20075	0.20296
0.3	0.29961	0.29943	0.29928	0.29921	0.29915	0.29675	0.30117	0.30447
0.4	0.39954	0.39931	0.39915	0.39904	0.39900	0.39577	0.40163	0.40602
0.5	0.49951	0.49925	0.49907	0.49898	0.49891	0.49484	0.50214	0.50760
0.6	0.59949	0.59923	0.59905	0.59893	0.59888	0.59395	0.60269	0.60919
0.7	0.69950	0.69926	0.69906	0.69895	0.69890	0.69310	0.70326	0.71082
0.8	0.79954	0.79932	0.79915	0.79907	0.79901	0.79231	0.80389	0.81248
0.9	0.89965	0.89943	0.89929	0.89922	0.89915	0.89157	0.90453	0.91415
	System: $(1-x)$ SnTe–$x TeI_4$					Sample TeI_4 $x = 1/\Phi$		
0.1	0.09986	0.09979	0.09975	0.09972	0.09970	0.09891	0.10037	0.10148
0.2	0.19978	0.19968	0.19960	0.19956	0.19953	0.19791	0.20084	0.20303
0.3	0.29975	0.29962	0.29954	0.29948	0.29945	0.29697	0.30135	0.30459
0.4	0.39975	0.39961	0.39952	0.39946	0.39943	0.39609	0.40189	0.40621
0.5	0.49975	0.49962	0.49952	0.49947	0.49946	0.49524	0.50247	0.50783
0.6	0.59979	0.59966	0.59959	0.59955	0.59952	0.59442	0.60307	0.60946
0.7	0.69981	0.69973	0.69966	0.69961	0.69961	0.69363	0.70369	0.71113
0.8	0.79988	0.79981	0.79975	0.79973	0.79972	0.79285	0.80432	0.81280
0.9	0.89995	0.89991	0.89987	0.89986	0.89988	0.89210	0.90495	0.91445
	System: $(1-x)$ PbTe–$x TeI_4$					Sample TeI_4 $x = 1/\Phi$		
0.1	0.10025	0.10039	0.10049	0.10058	0.10063	0.09958	0.10090	0.10186
0.2	0.20042	0.20063	0.20079	0.20092	0.20101	0.19899	0.20168	0.20363
0.3	0.30051	0.30076	0.30095	0.30109	0.30119	0.29827	0.30235	0.30533
0.4	0.40054	0.40079	0.40099	0.40115	0.40125	0.39744	0.40295	0.40697
0.5	0.50053	0.50077	0.50096	0.50110	0.50119	0.49655	0.50348	0.50856
0.6	0.60047	0.60070	0.60085	0.60098	0.60105	0.59558	0.60396	0.61011
0.7	0.70039	0.70055	0.70069	0.70080	0.70087	0.69459	0.70443	0.71166
0.8	0.80026	0.80039	0.80048	0.80057	0.80062	0.79352	0.80482	0.81318
0.9	0.90015	0.90020	0.90025	0.90028	0.90034	0.89243	0.90524	0.91465
	System: $(1-x)$ PbTe–xTeI					Sample TeI_4 $x = 4/\Phi$		
0.1	0.10029	0.10045	0.10055	0.10064	0.10072	0.09964	0.10094	0.10190
0.2	0.20051	0.20079	0.20099	0.20115	0.20128	0.19917	0.20181	0.20374
0.3	0.30068	0.30102	0.30130	0.30150	0.30160	0.29858	0.30259	0.30551
0.4	0.40077	0.40117	0.40148	0.40171	0.40188	0.39789	0.40328	0.40720
0.5	0.50080	0.50121	0.50151	0.50174	0.50191	0.49707	0.50387	0.50886
0.6	0.60076	0.60113	0.60143	0.60161	0.60177	0.59610	0.60437	0.61043
0.7	0.70073	0.70094	0.70115	0.70131	0.70144	0.69501	0.70476	0.71190
0.8	0.80043	0.80062	0.80076	0.80086	0.80094	0.79376	0.80503	0.81331
0.9	0.90017	0.90017	0.90021	0.90023	0.90026	0.89240	0.90518	0.91463
	System: $(1-x)$ PbTe–$x PbI_2$					Sample TeI_4 $x = 2/2 + \Phi$		
0.1	0.10027	0.10042	0.10054	0.10063	0.10069	0.09971	0.10088	0.10172
0.2	0.20048	0.20072	0.20094	0.20109	0.20121	0.19946	0.20155	0.20304
0.3	0.30061	0.30094	0.30120	0.30140	0.30154	0.29927	0.30201	0.30397
0.4	0.40068	0.40105	0.40135	0.40156	0.40171	0.39914	0.40228	0.40452
0.5	0.50070	0.50108	0.50136	0.50159	0.50175	0.49907	0.50234	0.50467
0.6	0.60066	0.60100	0.60129	0.60149	0.60164	0.59909	0.60222	0.60446
0.7	0.70056	0.70086	0.70109	0.70127	0.70140	0.69918	0.70193	0.70389
0.8	0.80042	0.80069	0.80082	0.80096	0.80105	0.79937	0.80145	0.80295
0.9	0.90024	0.90034	0.90045	0.90053	0.90056	0.89963	0.90082	0.90166

Table 2 (continued)

1	2	3	4	5	6	7	8	9
System: (1 − x) PbTe–xPbI					Sample TeI_4 $x = 4/4 + \Phi$			
0.1	0.10028	0.10044	0.10056	0.10066	0.10072	0.09972	0.10090	0.10173
0.2	0.20052	0.20080	0.20103	0.20119	0.20132	0.19953	0.20159	0.20308
0.3	0.30068	0.30106	0.30136	0.30158	0.30174	0.29940	0.30211	0.30404
0.4	0.40079	0.40122	0.40157	0.40184	0.40203	0.39934	0.40242	0.40462
0.5	0.50083	0.50129	0.50166	0.50193	0.50213	0.49933	0.50254	0.50482
0.6	0.60081	0.60125	0.60162	0.60188	0.60206	0.59938	0.60243	0.60363
0.7	0.70072	0.70111	0.70142	0.70165	0.70183	0.69946	0.70215	0.70405
0.8	0.80055	0.80085	0.80109	0.80128	0.80140	0.79961	0.80165	0.80308
0.9	0.90030	0.90049	0.90061	0.90072	0.90080	0.89978	0.90093	0.90173
System: (1 − x) PbTe–xI					Sample TeI_4 $x = 4/4 + \Phi$			
0.1	0.10028	0.10041	0.10053	0.10062	0.10069	0.09970	0.10087	0.10172
0.2	0.20045	0.20071	0.20090	0.20105	0.20115	0.19943	0.20153	0.20302
0.3	0.30056	0.30087	0.30111	0.30129	0.30141	0.29918	0.30194	0.30392
0.4	0.40060	0.40092	0.40117	0.41135	0.40148	0.39898	0.40216	0.40444
0.5	0.50057	0.50087	0.50110	0.50127	0.50140	0.49883	0.50216	0.50455
0.6	0.60049	0.60074	0.60093	0.60107	0.60117	0.59876	0.60198	0.60430
0.7	0.70036	0.70054	0.70068	0.70077	0.70085	0.69880	0.70164	0.70369
0.8	0.80020	0.80031	0.80039	0.80044	0.80048	0.79896	0.80117	0.80274
0.9	0.90007	0.90011	0.90013	0.90015	0.90017	0.89934	0.90059	0.90149

recorded after Reed-Long. The programme has entirely been recorded on FORTRAN-4 after an algorithm close to the one offered by ANDRUSHCHENKO.

Table 2 shows the results from the calculations of X along different hypothetical section of the systems Pb–Te–I and Sn–Te–I at different values of θ and at fixed value of the accelerating voltage (from 15 to 35 kV, at intervals of 5 kV), and at a fixed value $\theta = 18°$. IL_α and TeL_α were used as analytical lines.

It becomes clear from Tables 1 and 2 that, notwithstanding the remarkable changes in the relative intensities of the analytical lines IL_α and TeL_α, Φ_I^{Te} is slightly dependent on the variations of the conditions of analysis (value of the accelerating voltage U_0 and the angle of selection of the X-ray radiation (θ)).

It follows that, unlike the ZAF method which is inapplicable to the analysis of relief surfaces, the method of relative intensities is a promising one for holding such investigations. This method is also promising for the analysis of thin samples with variable thickness d, and it does not require mathematical processing of the experimental results with the use of computer. It must be added that, as shown by MOSHNIKOV, upon applicability of the method of relative intensities, the results from the calculations of Φ_j^i after different algorithms, even considerably differing from one another, yield similar results. This is explained by the fact that even in cases of considerable errors in the calculations of the corrective coefficients for the i-th and j-th elements, the errors are close in magnitude for the elements of close successive numbers and are compensated at the calculation of Φ_j^i.

For the application of the method of relative intensities it is necessary to know the section along which the change in the composition takes place, e.g. (1 − x) SnTe–$x$$SnI_2$ or (1 − x) SnTe–$x$$SnI_4$, etc.

The section along which the x composition is changing can be determined on control flat sections by the method of the isohypses, as proposed by ZASLAVSKY et al. The data necessary for analysis of the Sn–Te–I system have been examined by ASSENOV et al.

In this manner, the method of relative intensities proves to be a promising one in the X-ray spectral microanalysis of the tellurides of tin and lead, obtained by the iodide method, and ensures microanalysis of the composition also under conditions of changes in the values of the accelerating voltage U_0 and the angle of selection of the X-ray radiation θ.

References

ANDRUSHENKO, N. S.: in: Proceedings Aparatura i metodi rentgenovskogo analiza, No 5, 1975, p. 179 Leningrad, USSR

ASSENOV, R., MOSHNIKOV, V. A., YASKOV, D. A.: phys. stat. sol. (a) **88**, K27 (1985)

ASSENOV, R., MOSHNIKOV, V. A., YASKOV, D. A.: Cryst. Res. Technol. **21**, 1549 (1986)

ASSENOV, R., MOSHNIKOV, V. A., YASKOV, D. A.: Cryst. Res. Technol. **21**, 1553 (1986)

ASSENOV, R., MOSHNIKOV, V. A., PATAROV, B., YASKOV, D. A.: Cryst. Res. Technol. **21**, 1487 (1986)

DEDEGKAEV, T. T., MOSHNIKOV, V. A., YASKOV, D. A.: in: Fizicheskie metodi issledovanija tverdogo tela, Sverdlovsk 1982, p. 124—128

DEDEGKAEV, T. T., MOSHNIKOV, V. A.: in: Izvestija (LETI) Leningrad No 263, 1980 p. 100—108

DEDEGKAEV, T. T., MOSHNIKOV, V. A., YASKOV, D. A.: in: Proceedings Poluchenija i svoistva tonkih plenok Kiev 1981, p. 75—78

GAFNI, G., RAFAELOFF, R., SHAMIR, I.: Mat. Res. Bull. **14**, 217 (1979)

HEINRICH, K. F. I.: The absorption Correction Model for Microprobe Analysis — 2nd Nat. Conf. on Electron Microprobe Analysis, Boston 1967, p. 13

KWESTROO, W.: in: "Proceedings Preparativnie metodi v khimii tverdogo tela, P. HAGENMULLER — Ed. Moscow, 1976, p. 592

MOLDOVANOVA, M., TRIFONOVA, E. P., KARAGIOZOV, L., ASSENOV, R.: Bulgarian Author's Certificate Appl. No 42971 (21. 03. 1979)

MOLDOVANOVA, M., TRIFONOVA, E. P., ASSENOV, R., KARAGIOZOV, L.: phys. stat. sol., (a) **58**, (1980)

MOSHNIKOV, V. A.: Summary of the Ph. D. Thesis Issledovanie neodnorodnostey v tverdikh rastvorakh tellurida svintza-tellurida olova, Electrotechnical Institute of Leningrad, 1981

PHILIBERT, I.: Metaux, Corrosion Industries N 465, 157—194 (1964); N 466 pp. 216—268 (1964); N 469 p. 325—407 (1965)

REED, S. I. B.: Electronno-sondoviy microanalysis, Moscow 1979

REED, S. J. B., LONG, J. V. F.: Electron probe measurements near phase boundaries, in: X-ray Optics and X-ray Microanalysis, New York 1963 p. 317—327

ZASLAVSKIY, A. I., DEDEGKAEV, T. T., SIDOROV, A. F.: in: Proceeding Aparatura i metodi rentgenovskogo analiza No 5, 1969 p. 163, Leningrad, USSR.

(Received February 12, 1987)

Authors' address:

c/o Dr. R. ASSENOV
Semiconductor Physics and Technology Research Laboratory, University of Sofia
Boulvd "A. Ivanov" No 5
Sofia 1126, Bulgaria

Cryst. Res. Technol.	22	1987	10	1297–1300

S. C. JAIN, H. K. SINGH, K. S. BALAKRISHNAN, S. CHANDRA

National Physical Laboratory, New Delhi, India

A Simple Optical Method of Measuring the Tilt Bias Angle of Liquid Crystals

A simple optical method based on refractive indices measurement for determining the tilt bias angle of liquid crystals is described. The required refractive indices n_o, n_e and n_θ are measured by (i) Abbe's refractometer and by (ii) optical wedge method. The tilt bias angle by this technique can be determined to accuracy of $\pm 0.2°$ in the high tilt angle range (20° to 40°). This method is rather simple and quick and does not require any external magnetic field and/or knowledge of other physical constant of liquid crystals, as are necessary in some of the earlier reported methods.

Es wird eine optische Methode zur Bestimmung des tilt-bias Winkels von Flüssigkristallen beschrieben, die auf der Messung von Brechungsindizes basiert. Die erforderlichen Brechungsindizes n_o, n_e und n_θ wurden 1. mittels Abbé-Refraktometer und 2. mit Hilfe der Prismenmethode bestimmt. Die auf diese Weise ermittelten Winkel haben im Bereich von 20–40° einen Fehler von $\pm 0{,}2°$.
Im Unterschied zu anderen Methoden sind so einfache und schnelle Messungen ohne den Einsatz äußerer Magnetfelder und ohne Kenntnis weiterer physikalischer Konstanten der Flüssigkristalle möglich.

Liquid crystal multiplexed display devices which have come into great prominence during the last decade, depend critically for their electro-optic performance characteristics on the tilt bias angle (GOODMAN et al.) (the angle which the uniformly aligned liquid crystal (LC) molecules make at the bounding glass surface). The effect of the high tilt angle is also extremely important in recently discovered bistable displays based on super twisted birefringence effect (SBE) (SCHEFFER, NEHRING 1984; VAN SPRANG et al.). It is therefore, necessary to measure it accurately.

Several methods (MEYERHOFER; KAHN; SCHEFFER, NEHRING 1978; TORIYAMA, ISHIBASHI; BAUR et al.; NAKANO et al.; CROSSLAND et al.; URBACH et al.; LEVY et al.; KOMITOV et al.) have been proposed so far of measuring the tilt bias angle of liquid crystals and they are based on either capacitance measurement (MEYERHOFER; KAHN; SCHEFFER, NEHRING 1978; TORIYAMA, TSHIBASHI) or on optical phase retardation measurement (BAUR et al.; NAKANO et al.; CROSSLAND et al.; URBACH et al.; LEVY et al.; KOMITOV et al.) with or without external field and are called destructive or non-destructive, respectively. Most of these methods are quite intricate, destructive and take a fairly long time in measurements.

A very simple, accurate and quick method based on the refractive indices measurement is proposed herewith. It is well known that the nematic liquid crystals are birefringent and when a linearly polarized beam of light incidents on a thin uniformly aligned nematic slab with its $\boldsymbol{E}$ vector either parallel or perpendicular to the preferred director's direction, it experiences refractive index n_e or n_o, respectively. However, when the LC molecules make a finite angle (tilt bias angle) with the bounding glass surfaces, the polarized light with its $\boldsymbol{E}$ vector making an angle θ with the liquid crystal molecules will experience a refractive index n_θ that varies with θ as:

$$n_\theta = n_e \left(1 + \frac{n_e^2 - n_o^2}{n_o^2} \cdot \sin^2 \theta\right)^{-1/2} \tag{1}$$

and the polarized light with its $\boldsymbol{E}$ vector normal to the LC molecules will still experience the refractive index n_0. So an accurate measure of n_e, n_θ and n_o will enable to determine the tilt bias angle θ accurately.

The refractive index n_o can be easily measured with Abbe's refractometer and n_e and n_θ can be determined from the birefringence measurements using the modified optical wedge method (JAIN). In this method, a uniformly aligned nematic layer is sandwiched between two optically flat glass plates and the wedge cell is formed by placing a mylar strip (50—100 μm thick) at one of the two edges of the cell (Fig. 1). The cell is only partly filled with the liquid crystalline material and when it is viewed in the reflection geometry at normal incident linearly polarized light with the polarizing axis making an angle of 45° with the edge of the wedge, one observes equi-distant interference fringes due to birefringence in the liquid crystal region. The distance ΔX between the successive fringes is related to the wedge angle ψ, wavelength of light λ and the birefringence $\Delta n = n_e - n_0$ as:

$$\Delta X = \frac{\lambda}{2\psi \cdot \Delta n} \tag{2}$$

$$(\tan \psi = \psi, \text{ when } \psi \text{ is small})$$

provided the LC molecules are homogeneously (planar) aligned in the planes of the two glass plates.

When the same cell is viewed in the air gap region, one observes equidistant Fizeau fringes with the spacing between the two successive fringes given by:

$$\Delta X' = \frac{\lambda}{2\psi}. \tag{3}$$

In the case when the LC molecules instead of lying in the planes of the two glass plates make a finite angle θ with them, then the interference fringes will be caused

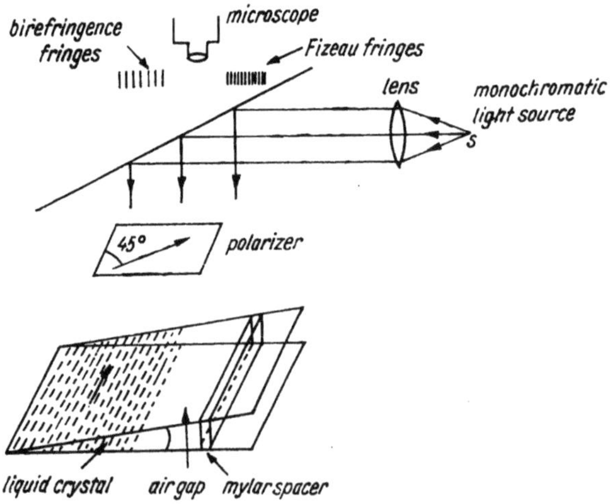

Fig. 1. Basic experimental set up of the optical wedge method for measuring birefringence of liquid crystals

by the effective birefringence $\Delta n_\theta = n_\theta - n_o$ and not by $\Delta n = n_e - n_o$. So in the tilted sample, the interference fringe width $\Delta X''$ will correspond to:

$$\Delta X'' = \frac{\lambda}{2\psi \, \Delta n_\theta} . \tag{4}$$

From eqs. (2)—(4), Δn and Δn_θ and in turn n_e and n_o can be easily determined. Thus, by knowing n_0, n_e and n_θ; θ can be found from eq. (1) as:

$$\sin \theta = \frac{n_o}{n_e} \left(\frac{n_e^2 - n_\theta^2}{n_e^2 - n_0^2} \right)^{1/2} . \tag{5}$$

The fringe width in the present measurements has been measured with a comparator to an accuracy of 0.0001 cm and the birefringence which is independent on the measure of the wedge angle, is measured to an accuracy of 0.5%.

All the measurements in the present investigations were done on La-Roche mixture RO-TN-403 at room temperature (26.7 ± 0.2 °C) to avoid any variation in refractive indices due to temperature. The zero tilt bias cell to determine Δn and n_e was made by treating the glass plates with 1% solution of PVA followed by unidirectional rubbing. The homogeneous alignment at the lass surfaces was also produced by obliquely depositing SiO under high vacuum ($p \leqq 10^{-5}$ torr). The deposition angle of SiO controlled the tilt bias angle 'θ' of the LC molecules.

Figure 2 is a typical plot of the measured tilt bias angle θ versus the deposition angel α of the SiO (α is the angle between the normal to the substrate and the evaporation direction). The tilt bias angle is $\approx 0°$ for the deposition angle $55° < \alpha$ 65° and is $> 20°$ for $\alpha > 75°$. The observed tilt angle dependence on the SiO evaporation angle is in good agreement with the previously reported results (UCHIDA et al.). The accuracy in the tilt bias angle measured is ±0.2° for tilt bias angle > 20° and is within 1° for tilt bias angles < 5°. However, the tilt bias angle differed considerably from sample to sample at the higher angles of deposition. Therefore, we investigated the tilt bias angle dependence on the deposition parameters, namely (i) the deposition rate and (ii) the thickness of SiO coating.

It is observed that the tilt bias angle varied significantly with the deposition rate of SiO. Figure 3 shows that the tilt bias angle varied from 2.8 ± 2° to 36 ± 2° when the deposition rate varied from ≧ 30 Å/minute to ≦ 5 Å minute at 85° SiO deposition angle. Similar variations in the tilt bias angles were also observed at other deposition

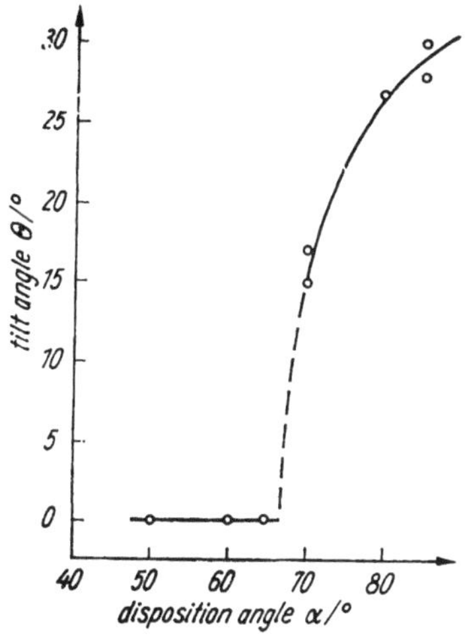

Fig. 2. Plot of the tilt angle θ versus the deposition angle α of SiO. RO-TN-403. Thickness 200 Å; evaporation rate ≧ 20 Å · min^{-1}

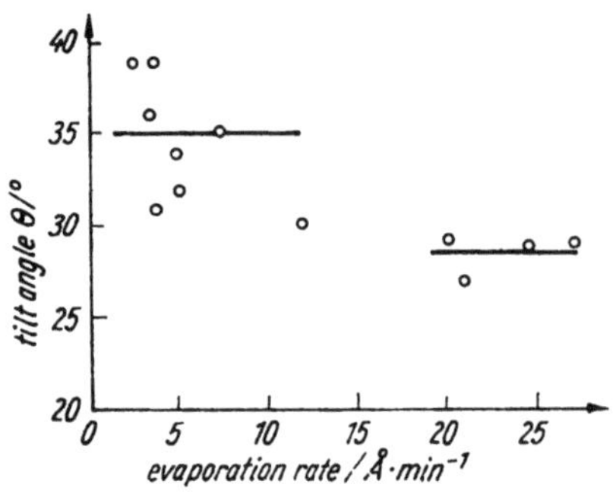

Fig. 3. Variation of the tilt angle θ with the deposition rate. $\alpha = 85°$; $p = 10^{-5}$ torr. RO-TN-403

angles (80°, 70°, and 65°). However, little effect of deposition rate was observed at deposition angles $\leq 60°$. These results are in some contradiction to the reported insensitivity of the tilt bias angle on deposition rate for all angles of deposition by UCHIDA et al. who had used deposition rates ≥ 150 Å/minute. We have, however, used extremely slow deposition rates and might be mainly responsible for the observed dependence of the tilt bias angle on the deposition rate of SiO.

We have also studied the effect of SiO coating thickness on the tilt bias angle. It is observed that SiO thickness has little effect on the tilt bias angle provided the deposition rate was maintained constant. The variation in the tilt bias angle was $\leq 2°$ for SiO (85°) coating between 50 and 300 Å.

It is concluded from the above studies that the optical wedge method for the birefringence measurements along with Abbé's refractometer can be conveniently used to determine the tilt bias angle accurately. It has also emerged through by studies that the rate of deposition of SiO affects the tilt bias angle considerably.

The authors are highly thankful to Director, N. P. L. for his constant encouragement and interest in this work.

References

BAUR, G., WITTER, V., BERREMAN, D. W.: Phys. Lett. **A 56**, 142 (1976)
CROSSLAND, W., MORRISSY, J., NEEDHAM, B.: J. Phys. **D9**, 2001 (1976)
GOODMAN, L. A., MEYERHOFFER, D., DIGIOVANNI, S.: IEEE Trans. Electron Devices **ED-23**, 1176 (1976)
JAIN, S. C.: Ph. D Thesis, Delhi University 1981
KAHN, F.: Mol. Cryst. Liq. Cryst.: **38**, 109 (1977)
KOMITOV, L., HAUCK, G., KOSWIG, H. D.: Cryst. Res. Technol. **19**, 253 (1984)
LEVY, Y., RIVIERE, D., IMBERT, C., BOIX, M.: Opt. Commun. **26**, 225 (1978)
MEYERHOFFER, D.: Phys. Lett. **A51**, 407 (1975)
NAKANO, F., ISOGANI, M., SATO, M.: Jpn. J. Appl. Physics **19**, 2013 (1980)
SCHEFFER, T. J., NEHRING, J.: J. Appl. Physics **48**, 1783 (1978)
SCHEFFER, T. J., NEHRING, J.: Appl. Phys. Lett. **45**, 1021 (1984)
SPRANG, H. A. VAN, AARTSEN, R. G., DE VAAN, A. S. M.: J. Appl. Physics **59**, 3087 (1986)
TORIYAMA, K., ISHIBASHI, T.: Non-Emissive Electrooptic Displays, edited by A. R. KMETZ and F. K. VON WILLISEN, New York 1976, p. 145
UCHIDA, T., OHGAWARA, M., WADA, M.: Jpn. J. Appl. Physics **19**, 2127 (1980)
URBACH, W., BOIX, M., GUYONZ, E.: Appl. Phys. Lett. **25**, 479 (1974)

(Received March 3, 1987)

Authors' address:

Dr. S. C. JAIN, H. K. SINGH, Mrs. K. S. BALAKRISHNAN, S. CHANDRA
National Physical Laboratory
Hillside Road
New Delhi 110012, India

Cryst. Res. Technol.	22	1987	10	1301–1311

V. VALVODA, R. ČERNÝ, R. KUŽEL jr., L. DOBIÁŠOVÁ

Faculty of Mathematics and Physics, Charles University, Prague, Czechoslovakia

Method for Complex X-ray Diffraction Analysis of TiN Coatings

A complex of X-ray diffraction methods for study of polycrystalline thin layers is presented. It is based on the approximation of measured profiles by suitable analytical functions. An application to TiN coatings prepared by PVD method on high-speed steel is presented. The necessity to investigate several diffraction parameters (position, integral intensity, breadth, asymmetry) of more reflections is emphasized.

Einige röntgenographischen Methoden für das Studium polykristalliner dünner Schichten werden vorgestellt. Sie basieren auf der Approximation der gemessenen Profile durch passende analytische Funktionen. Die Anwendung für die mit der PVD-Methode hergestellten TiN-Schichten auf Schnellschnittstählen wird gezeigt. Die Notwendigkeit der Untersuchung von mehreren Diffraktionsparametern (Position, Integralintensität, Breite, Asymmetrie) für mehrere Reflexionen wird betont.

1. Introduction

Thin layers of TiN and other compounds with excellent mechanical properties play evergrowing role as protective coatings. For this reason the need of complete understanding of their microstructure have arisen in the last time. The unique character of diffraction methods and their specialities for thin layers were described elsewhere (HUANG, PARRISH). For understanding the phenomena in X-ray diffraction analysis of thin layers it is inevitable to use more complementary diffraction methods. The main complication arises from overlapping of broad profiles of TiN layer and Fe substrate. The best solution is the "Total Pattern Analysis" (LANGFORD et al.) based on the approximation of measured profiles by some analytical functions. In this paper we describe, using this base, the complex of diffraction methods specified for the case of TiN layers and allowing the new way of profile analysis. Some useful formulas are also collected here for practical applications.

2. Symbols

ω, θ, ψ	angles of rotation around ω-axis of goniometer
θ_{hkl}	Bragg angle of the reflection *hkl*
μ	linear absorption coefficient
t	layer thickness
a_0	"true" lattice parameter
σ	stress
$\alpha = \alpha' + \alpha''$	probability of single α' and double α'' deformation stacking faults
β	probability of growth stacking fault
G, V, X, Γ	parameters related to stacking faults and stress

S_1, S_2	elastic constants of a polycrystalline material
d_{hkl}	spacing of crystallographic planes (hkl)
d_0	interplanar spacing of unstressed layer
E, ν	Young modulus, Poisson ratio
B	integral breadth of diffraction profile
e	strain in a layer
λ	X-ray wavelength
D	size of coherently diffracting domains
K	Scherrer's shape parameter (~ 1)
$2\theta_{hkl}^{C}$, $2\theta_{hkl}^{M}$	centroid, maximum of reflection hkl
m_{hkl}, F_{hkl}	multiplicity, structure amplitude of the reflection hkl
N_{hkl}	volume fraction contributing to intensity of reflection hkl
FWHM	full width at half maximum
x	ratio of molar concentration of N and Ti atoms

3. Experimental methods

TiN coating 1—10 μm thick were prepared by reactive sputtering in planar magnetron on the high-speed steel substrate at temperatures 50—150 °C. The diffraction experiment was carried out on the powder diffractometer allowing the independent motion of θ and 2θ circles. The CuKα radiation, we have used, is advantageous for the

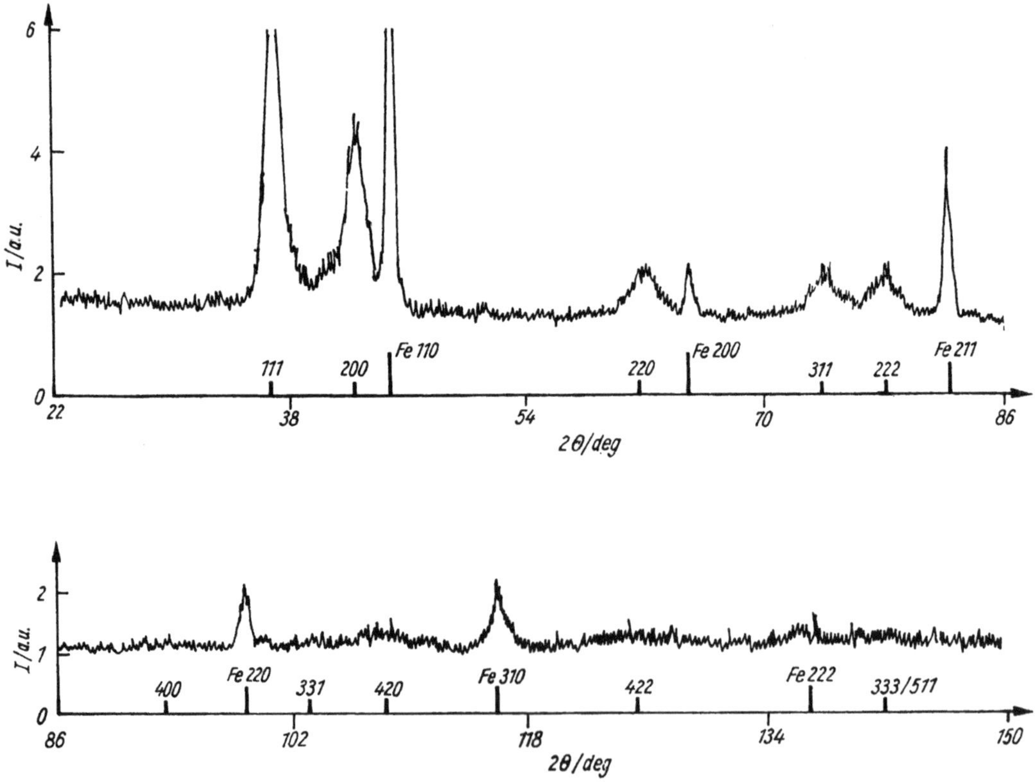

Fig. 1. Continuous diffraction record of a typical sample of TiN coating on HSS substrate. Reflections indicated with indices only belong to TiN

TiN and Fe reflections intensity ratio. The graphite monochromator in diffracted beam improves the peak to background ratio. The using of Soller slits is inevitable for lowering the instrumental contribution to profile asymmetry.

As a first step each sample was characterized by a continuous scan (Fig. 1), from which it was possible to determine parts containing 2—4 reflections, which were then thoroughly measured with step scan. As the main problem in diffraction analysis of thin layer arises from the profiles overlapping, we have performed the computer separation of profile clusters by fitting of analytical functions. In the same way the separation of the X-ray CuK_α doublet was done and the profile parameters (position, integral intensity, breadth, shape parameter) were obtained. We have used the computer program PROF, the modified version of PROFIT (VALVODA, VOLAND). It uses Levenberg-Marquardt least squares method and enables to choose from 5 types of functions: Pearson VII, pseudo-Voigt and their asymmetric versions and rational function (Tab. 1). The examples of approximated diffraction profiles are presented in Figure 2.

The asymmetry of profile was characterized by (COHEN, WAGNER)

$$\Delta 2\theta_{hkl} = 2\theta_{hkl}^{C} - 2\theta_{hkl}^{M} \,. \tag{1}$$

Stress in the layer is given by the dependence of lattice distortion on the measuring direction (JAMES, COHEN). This macroscopic distorsion was studied by $\omega - 2\theta$ scan of the reflection 422 of TiN ($\sin^2 \psi$ method). The measured profile (Fig. 3) was corrected for Lorentz-polarization factor and absorption factor for the case of thin layer (PEISER et al.), which also includes the change of irradiated volume with ψ

$$A = \frac{1}{2\mu} (1 + \tan \psi \cdot \cot \theta) \left[1 - \exp \left(-2\mu \frac{\sin \theta \cdot \cos \psi}{\sin^2 \theta \cos^2 \psi - \cos^2 \theta \sin^2 \psi} t \right) \right]. \tag{2}$$

In the case of broad profiles it is not necessary to correct the data for background and α_2 contribution (JAMES, COHEN). Position of profile maximum is to a good approximation given by a parabola fitted to the points laying above 80% of the maximum.

Table 1

Analytical functions used in the program PROF. y is number of pulses per second after correction for Lorentz-polarization factor, $x = \sin \theta$

Parameters: $A1$ ... height of maximum
$A2$ ... position of maximum
$A3$, $A5$ ... parameters connected with line breadth
$A4$ ... shape parameter

1. Modified Lorentz function (Pearson VII):

   $y = A1/(1 + A3(x - A2)^2)^{A4}$

2. Pseudo-Voigt function:

   $y = A1[A4/(1 + A3(x - A2)^2) + (1 - A4) \exp(-A5(x - A2)^2)]$

   It is often sufficient to use $A3 = A5$

3. Rational function:

   $y = A1/(1 + A3(x - A2)^2 + A5(x - A2)^4)$

4. and 5. Asymmetrical versions of 1. and 2.:

   For $(x - A2) < 0$ $A3$ and $A4$ are substituted with $A6 \cdot A3$ and $A7 \cdot A4$.

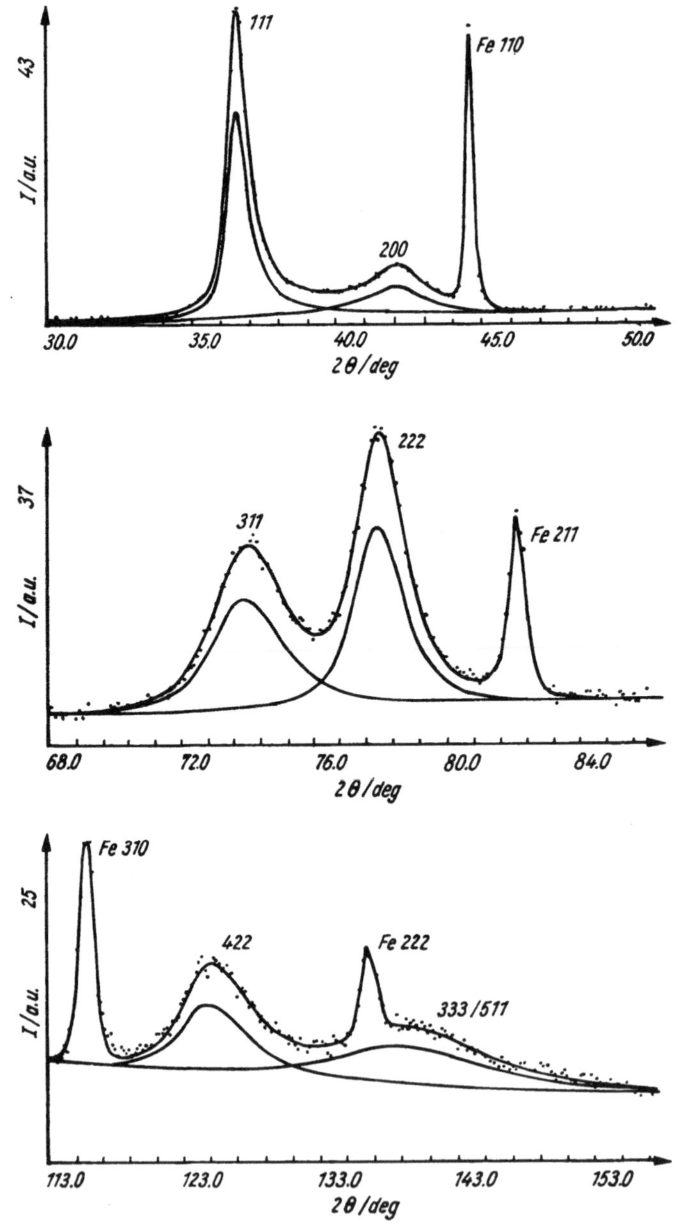

Fig. 2. Step scan measurements approximated with analytical functions. The α_1 components of CuK_α doublet are drawn for TiN only

The characterization of preferred orientation of grains was performed using the FWHM of ω-scan of a suitable reflection. In Figure 4 you can see an example and also the connection between the angles of rotation is clearly visible

$$\psi = \theta_{hkl} - \omega\,. \tag{3}$$

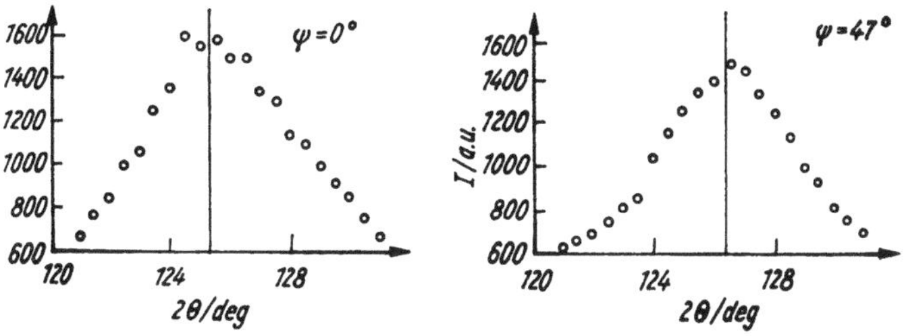

Fig. 3. Step scan measurement of the 422 reflection of TiN for two values of sample inclination angle ψ in stress determination

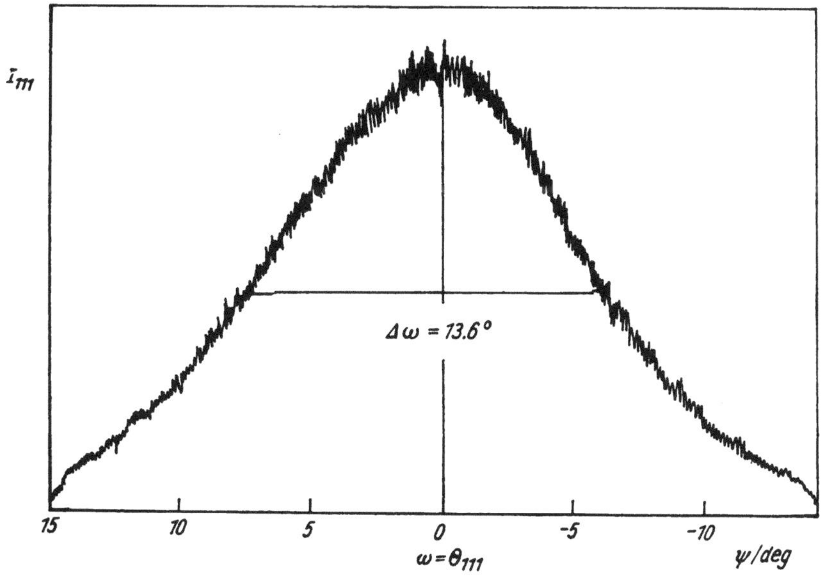

Fig. 4. Continuous ω-scan measurement of the 111 reflection of TiN. FWHM is given by the horitzonal bar

The measured curve must be again corrected for the absorption effect (2). A defocusation of the diffracted beam in the ω or $\omega - 2\theta$ scan must be minimized using a narrow divergence slit and a broad receiving slit.

4. Evaluating methods

Lattice parameters were calculated from the positions of individual reflections corrected for goniometer misalignement using external standard-powder tungsten carbide. These parameters are usually described (WAGNER) by the expression

$$a_{hkl} = a_0 + S_1(hkl)\, a_0\sigma + \alpha a_0 G(hkl) + s \cdot \cot \theta_{hkl} \cdot \cos \theta_{hkl} \,, \tag{4}$$

where the second term represents the influence of stress, the third the influence of rowthg stacking faults and the fourth the instrumental correction. The elastic constants $S_1(hkl)$ can be approximated by bulk moduli

$$S_1 = -\nu/E \,, \tag{5}$$

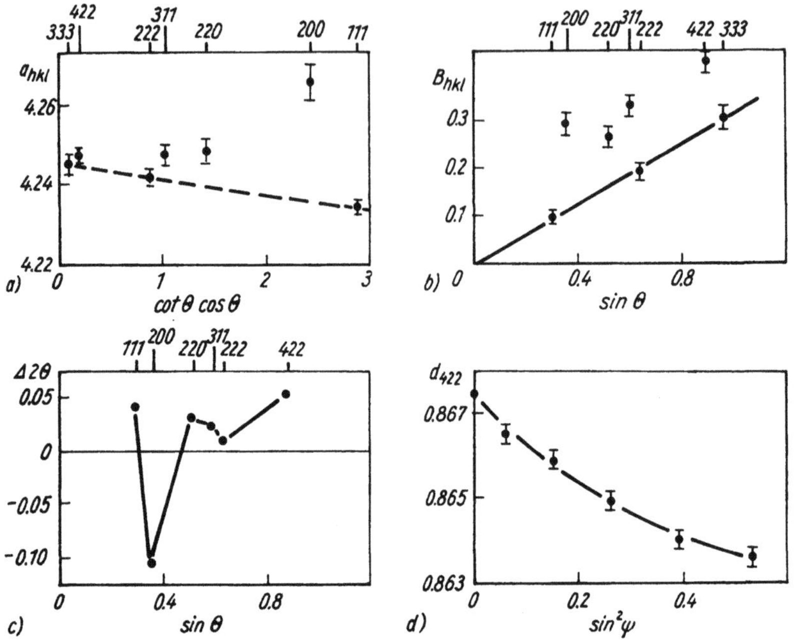

Fig. 5. Characteristic parameters of a TiN coating obtainable with diffraction methods: a) lattice parameters a_{hkl} in unit of 10^{-10} m, b) line breadth B_{hkl} in units of nm^{-1}, c) asymmetry of line profiles in units of nm^{-1}, d) spacing of diffraction planes (422) d_{422} in units of 10^{-10} m. Reflection indices are indicated in a), b), and c)

or measured (MARION, COHEN a) or calculated as an average from the Reuss $S_1^R(\Gamma_{(hkl)})$ and the Voigt S_1^V approximations (WAGNER). The measured values of a_{hkl} can be fitted by a least squares method. However, when $S_1(hkl)$ is unknown, what was our case, the values of a_0, σ and α cannot be separated. It must be said that in most cases studied it was impossible to describe the observed scatter of lattice parameters by (4).

Stress measurements were carried out for two rotations of a sample around the surface normal differing by an angle of 180°. The mean values from these two measurements are given in Figure 5d. Providing the isotropy of stress in the plane of the layer and absence of stress normal component in the whole diffracting volume the measurement can be evaluated (JAMES, COHEN)

$$d_{hkl}(\psi) = d_0(hkl) \cdot \left(1/2 \cdot S_2(hkl)\,\sigma \cdot \sin^2\psi + S_1(hkl)\,\sigma + 1\right). \quad (6)$$

The elastic constants $S_2(hkl)$ can be obtained in the same manner as $S_1(hkl)$. We have used again the approximation of bulk moduli

$$S_2 = (1 + \nu)/E. \quad (7)$$

As d_0 we have used $d_{hkl}(\psi)$ for $\psi = 35.26°$ as follows from (HAASE) and from the value of $\nu = 0.2$ for bulk TiN. The dependence of d_{hkl} vs $\sin^2\varphi$ is presumed to be linear in (6). However, we have observed very often curved or oscillating dependence. If there is a correlation of this oscillations with texture, it is possible to obtain the correct value of σ (MARION, COHEN b). But the correct explanation is unknown at the present time.

The first approximative but relatively well consistent results from the line broadening can be obtained with integral breadths. Each such parameter must be corrected for an instrumental influence (DELHEZ et al. a; LANGFORD; KLUG, ALEXANDER). Using the integral breadths, we have chosen from broad spectrum of methods the correction employing the Voigt function (LANGFORD; DE KAIJSER et al.). It consists of the separation of the whole integral breath B on Cauchy B_C and Gauss B_G components, using the known dependence (LANDFORD) of B_C/B and B_G/B on the ratio of FWHM and integral breadth of a diffraction profile given by the experiment. Then it holds

$$\left.\begin{aligned} B_C^f &= B_C^h - B_C^g\,, \\ (B_G^f)^2 &= (B_G^h)^2 - (B_G^g)^2 \end{aligned}\right\} \tag{8}$$

where indices f, h and g indicate the physical, measured, and instrumental profile, respectively. The corrected components B_C^f and B_G^f can be again synthetized into the whole corrected integral breadth using the known dependences B/B_G vs B_C/B_G (LANGFORD).

Strain e (inhomogeneous relative changes of lattice spacing) and grains size D (more accurately coherent domains size) are the classical physical quantities related to the line broadening. It is often assumed that strain produces Gauss and grain size Cauchy profile shape, related to B_G^f and B_C^f respectively. Accepting this crude presumption, we can use the method of one reflection (LANGFORD).

Having measured more reflections (the best is more reflection orders from one set of planes), we can evaluate the data with the method of more reflections. It can be written (LANGFORD)

$$\left.\begin{aligned} B_C^f(hkl) &= K/D_C + (4e_C/\lambda)\cdot \sin\theta_{hkl}\,, \\ \left(B_G^f(hkl)\right)^2 &= (K/D_G)^2 + (4e_G/\lambda)^2\cdot \sin^2\theta_{hkl}\,. \end{aligned}\right\} \tag{9}$$

Indices indicate the individual components. The quantities D_G, D_C, e_G, e_C can be estimated from (9) and synthetized into total D, e (DELHEZ et al. a).

Taking into account the accumulation of experimental errors in work with line breadth components and that the strain is a dominant reason of line broadening (Fig. 5b), we have used the method of more reflections working with the total integral breadths B. In such a case, it is not crucial which of dependencies (9) will be used for evaluation (ALLAN).

The Fourier analysis of line profiles is most frequently connected with the classical model of Warren and Averbach. The good review of basic relations and contemporary state of methodology is in (DELHEZ et al. b).

The reliability of the parameter D depends on the real structure of the sample. We can found essentially three cases:

a) the values of D are the true sizes of grains,
b) the values of D are connected with the sizes of subgrains and they depend on the boundaries of these subgrains. In many cases the values of D do not have a meaningful interpretation (WILKENS),
c) the values of D are given by stacking faults. In this case they are dependent on the reflection used. For the f.c.c. structure it holds

$$1/D(hkl) = (1.5\alpha + \beta)/(2a_0)\cdot V(hkl)\,. \tag{10}$$

The interpretation of the strain e can be done fenomenologically or in connection with some lattice disorder distribution. In the first case we can write

$$e(hkl) = \sigma/E(hkl)\,, \tag{11}$$

or according to (ROTHMAN, COHEN) divide the strain into Reuss and Voigt components. Strain can be formally evaluated also by the theory of (KRIVOGLAZ).

The asymmetry of diffraction profile is for the f.c.c. structure connected with the probabilities of stacking faults on the (111) planes by (WAGNER)

$$\Delta 2\theta_{hkl} = 14.6\cdot X(hkl)\cdot \tan\theta_{hkl}\cdot(\beta + 4.5\alpha'')\,. \tag{12}$$

Table 2
Constants for stacking faults in (111) planes and stress in f.c.c. crystals

hkl	*G*	*V*	*X*	*Γ*
111	−0.035	0.43	0.75	0.33
200	0.069	1	−1	0
220	−0.035	0.71	0.25	0.25
311	0.013	0.45	0	0.16
222	0.017	0.43	−0.75	0.33
400	−0.035	1	1	0
331	−0.007	0.8		0.27
420	0.007	0.22		0.16
422	0	0.82		0.24
333/511	0.003	0.61		0.33/0.07

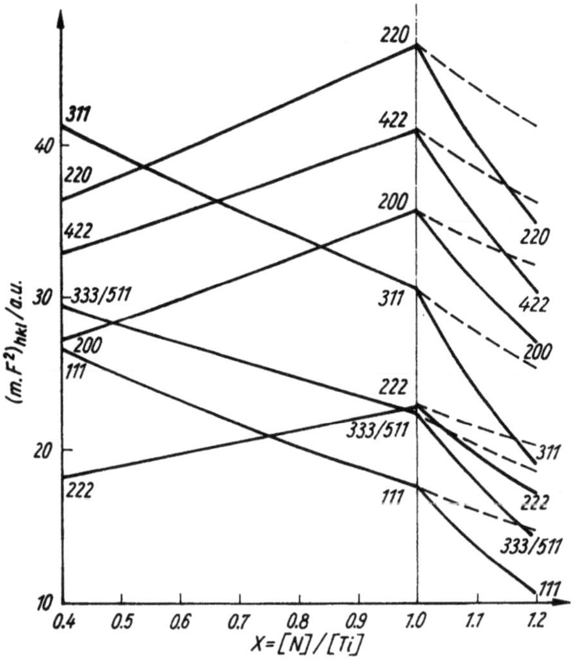

Fig. 6. Structure – dependent part of intensities of TiN reflections as a function of molar ratio of nitrogen and titanium. Overstoichiometric ratio can be realized by: vacancies on Ti sublattice (full line), substituting the N atoms instead of Ti atoms (dashed line)

The integral intensity of measured reflections can be after correction for the Lorentz-polarization factor written as

$$I_{hkl} = \text{const.}\ A(t, \theta_{hkl}, \psi) \cdot m_{hkl} \cdot F^2_{hkl} \cdot N_{hkl}(\psi)\ , \tag{13}$$

where $A(t, \theta_{hkl}, \psi)$ is the absorption factor (2). Summing the values of $N_{hkl}(\psi)$ over all measured reflections, we can characterize the degree of layer crystallinity. According to (ČAPKOVÁ et al.) it is possible to obtain an information about the composition x of TiN_x from the comparison of integral intensities of even and odd-type of reflections. Advantageous is to choose the reflections 111 and 222 which are influenced by preferred orientation in the same manner. In Figure 6 are shown calculated structure dependent parts of intensities ($m_{hkl} \cdot F^2_{hkl}$) of some reflections plotted as a function of the ratio $x = [N]/[Ti]$.

5. Results and discussion

Typical values of profile parameters and their errors are given in Table 3 and Figure 5. It is clear that the crystallographic anisotropy of lattice parameters and line breadths exceeds in these examples the range of experimental errors, which have been obtained from the least squares fitting. The errors of stress measurements (5d) has been taken from the mean value of two measurements differing in rotation of samples around the surface normal. The differences in stress values between different layers were also significantly greater than their errors. The integral intensities have been determined by the least squares fitting with the relatively highest error owing to the large broadening of generally weak reflections.

Table 3
Parameters of diffraction profiles and stress in a typical TiN coating and instrumental line breadths. (Errors given in parentheses correspond to the last digit.)

TiN					
hkl	lattice parameter nm	integral breadth nm^{-1}	integral intensity arb. u.	profile asymmetry nm^{-1}	instrumental broadening B^g nm^{-1}
111	.42556(6)	.129(3)	289(8)	.044	.0141(2)
200	.4286(2)	.36(1)	118(3)	—.05	.0140(2)
220	.4261(4)	.31(2)	43(1)	.05	.0138(2)
311	.426(1)	.30(2)	199(8)	.059	.0134(2)
222	.4253(1)	.24(1)	159(5)	.053	.0133(2)
422	.4258(1)	.40(2)	170(5)	.042	.0138(2)
333/511	.4264(1)	.41(4)	83(2)	—0.2	.0138(2)

Stress $\sigma = (-6.4 \pm 0.2)$ GPa

The most difficult problem in X-ray studies of thin films is to obtain a sufficient statistical precision in the step scan measurements. The value of 1000 pulses in the maximum of a reflection and a peak to background ratio 2/1 can be regarded as a minimum demand. As a criterion of computer fitting success we have used the factor

$$R = \sum_i \left(y(x_i) - I(x_i)\right)^2 / \sum_i I^2(x_i) , \qquad (14)$$

where $y(x_i)$ and $I(x_i)$ are fitted and measured values of intensity respectively. The obtained values of R were in the range .01—.035. However, the graphical visual control of fitted profiles is very needed in any case.

The question of a standard in the instrumental correction of the profiles is not critical in the case of very broad TiN lines. If it is impossible to use a suitable TiN standard, a different material with similar absorption (WC in our case) can be used. The nonideal standard correction (DELHEZ et al. a) can be applied in any case.

Prematured interpretation of the X-ray data obtained from TiN coatings often appears in the investigation of this material. Until a reliable interpretation of the observed anisotropy of lattice parameters is available, it will be at least courageous to determine the layer composition from single reflection measurement. A straightforward interpretation of the line broadening in terms of a grain size broadening only, without a detailed analysis, can be also often found in papers dealing with TiN. However, it can be seen from Figure 5b, that the major part of line broadening is the strain one. As well the stacking faults probability determined from the asymmetry of the 111 and 200 reflections only need not to be in accordance with asymmetry of other reflections.

The above presented complex of diffraction methods has been applied to 5 sets of TiN coatings differing in composition and thickness. The results are being prepared for publication.

6. Conclusions

Using the complex of diffraction methods it is possible to determine the following characteristic parameters of thin coatings:

1) lattice parameters
2) strain and grain size
3) stress
4) stacking faults probability
5) preferred orientation of grains
6) degree of crystallinity
7) stoichiometry.

The contribution of this method is in its complexity. An ordinary powder diffractometer and computer technique on a standard level is only needed. The approximation of the measured profiles by analytical functions solves the problem of overlapping lines and provides the easy way for obtaining the profile parameters.

The method has been applied on TiN layers but it is usable for arbitrary coatings giving measurable reflections on powder diffractometer.

Preliminary study of all obtainable microstructure parameters has shown the danger of premature conclusions based on individual parameters or on parameters derived from one reflection only.

References

Allan, C. G.: Improved graphical methods for separation of crystallite size and strain in X-ray diffraction analysis, U.K.A.E.A. TRG report 2548, Warrington 1974

Cohen, J. B., Wagner, C. N. J.: J. Appl. Physics **33**, 2073 (1962)

Čapková, P., Karen, P., Dobiášová, L.: Cryst. Res. Technol. **21**, 735 (1986)

Delhez, R., de Keijser, Th. H., Mittemeijer, E. J. (a): in: Proceedings Accuracy in powder diffraction, Maryland 1979, NBS publication 1980

Delhez, R., de Keijser, Th. H., Mittemeijer, E. J. (b): Fresenius Z. Anal. Chem. **312**, 1 (1982)

Haase, E. L.: Thin Solid Films **124**, 283 (1985)

Huang, T. C., Parrish, W.: Advances in X-ray analysis **22**, 43 (1978)

James, M. R., Cohen, J. B.: Treatise on materials science and technology **19A**, 1 (1980), ed. by Herbert Herman

de Keijser, Th. H., Langford, J. I., Mittemeijer, E. J., Vogels, A. B. P.: J. Appl. Cryst. **15**, 308 (1982)

Klug, H. P., Alexander, L. E.: X-ray diffraction procedures for polycrystalline and amorphous materials, New York 1970

Krivoglaz, M. A.: Difrakcija rentgenovskich lucej i nejtronov v neidealnych kristallach, Kiev 1983

LANGFORD, J. I.: in: Proceedings Accuracy in powder diffraction, Maryland 1979, NBS publication 1980
LANGFORD, J. I., LOUER, D., SONNEVELD, E. J., VISSER, J. W.: Powder Diffraction **1**, 211 (1986)
MARION, R. H., COHEN, J. B. (a): Advances in X-ray analysis **20**, 355 (1976)
MARION, R. H., COHEN, J. B. (b): Advances in X-ray analysis **18**, 466 (1974)
PEISER, H. S., ROOKSBY, H. P., WILSON, A. J. C.: X-ray diffraction by polycrystalline materials, Bristol 1960
ROTHMAN, R. L., COHEN, J. B.: Advances in X-ray analysis **12**, 208 (1969)
VALVODA, V., VOLAND, U.: Cryst. Res. Technol. **16**, 1197 (1981)
WAGNER, C. N. J.: in: Local atomic arrangements studied by X-ray diffraction (J. B. COHEN, J. E. HILLIARD, eds), New York 1966, p. 219
WILKENS, M.: phys. stat. sol. **A2**, 359 (1970)

(Received March 21, 1987)

Authors' address:

Dr. R. ČERNÝ, V. VALVODA, R. KUŽEL, jr., L. DOBIÁŠOVÁ
Faculty of Mathematics and Physics, Charles University
Ke Karlovu 5
121 16 Prague 2
Czechoslovakia

Book Review

G. L. Sewell

Quantum theory of collective phenomena

Clarendon Press, Oxford 1986, XIII + 229 p., £ 35.00, ISBN 0-19-851371-2

Zwischen endlichen und unendlich ausgedehnten Systemen gibt es grundlegende Unterschiede. Für letztere sind solche kollektiven Phänomene wie Phasenumwandlungen und irreversible Prozesse charakteristisch. Es ist verdienstvoll, daß die Besonderheiten solcher ausgedehnten Systeme ausgehend von den grundlegenden quantenmechanischen Formulierungen einmal im Zusammenhang und unter Einbeziehung modernster Gesichtspunkte der mathematischen Physik dargestellt wurden. Nach einer Einführung (Teil 1) werden im Teil 2 solche grundlegenden quantenmechanischen Konzeptionen ausgedehnter Systeme wie inäquivalente Darstellungen, lokale Observable, makroskopisch äquivalente Zustände, Stern-Algebren, das Gelfand-Naimark-Segal-Theorem, die Abwesenheit von Poincaré-Zyklen entwickelt. Danach wird in Teil 3 und 4 die statistische Thermodynamik behandelt: Konvexität und ihre Konsequenzen, Kontinuität und Semikontinuität, das Bolzano-Weierstraß-Theorem, thermodynamische Stabilitätsbedingungen, globale und lokale Stabilität, makroskopische Entartung, Symmetriebrechung, Kubo-Martin-Schwinger-Fluktuations-Dissipations-Bedingungen, Analytizitäts-, Singularitäts- und Konvexitäts/Konkavitäts-Eigenschaften des Helmholtzschen und des Gibbsschen Potentials, reine Phasen und Mischungen, thermodynamische Diskontinuität. Teil 5 befaßt sich mit Phasenumwandlungen: Ordnungsparameter, Ordnungs-Unordnungs-Übergänge (die Fröhlich-Simon-Spencer- und Peierls-Argumente beim klassischen Heisenberg-Ferromagneten bzw. Ising-Modell), kritische Phänomene (Landau-Theorie, kritische Singularitäten, Skalierungsgesetze, Renormierungstheorie, Dyson's hierarchisches Modell). Schließlich kommen im Teil 6 metastabile Zustände und im Teil 7 Ordnung und Phasenübergänge fernab vom Gleichgewicht (Dicke's Lasermodell, Fröhlich's Phononenmodell und Bose-Einstein-Kondensation) zur Sprache. Dem Autor gelingt es dankenswerterweise, Ergebnisse der mathematischen Physik in einer Weise zu verarbeiten, daß dem bemühten Leser der Zugang erleichtert wird. Dem dienen auch jeweils mehrere Anhänge zu jedem der genannten Teile.

Eine künftige noch umfassendere Behandlung kollektiver Phänomene wird auch solche wichtigen kooperativen Erscheinungen wie Supraleitung, Supraflüssigkeit, makroskopische und mesoskopische Quantenphänomene, Landausche Quasiteilchen u. ä. einbeziehen sollen.

Das Buch wendet sich an Physiker, Chemiker und Mathematiker, die an grundsätzlichen Fragen der Gleichgewichts- und Nichtgleichgewichtsstatistik der kondensierten Materie interessiert sind.

P. Ziesche

Cryst. Res. Technol.	22	1987	10	1313–1319

A. V. N. Gupta, J. V. Rao, P. Venkatacharyulu

Department of Physics, Nagarjuna University
Nagarjunanagar, India

Multifrequency Ultrasonic Investigations in the Liquid Crystal 5 OCB

The specific volume and ultrasonic velocity measurements (at four different fixed frequencies) are carried out in the liquid crystal 4'-n-pentyloxy-4-cyanobiphenyl (5 OCB). From these studies various physical parameters like adiabatic compressibility, molar sound velocity, molar compressibility, van der Waals' constant and intermolecular free lengths are estimated. The variation of the order parameter at the four frequencies is also studied. The results are analysed.

Introduction

The importance of the liquid crystalline phase transitions have been realized since the last two to three decades for their established technological applications in display devices and other manifold scientific uses. Just like as any ordinary liquid which exhibits phase transition, in any liquid crystalline compounds also which naturally exhibits phase transitions, any physical property like density, ultrasonic velocity, heats of transition etc. change abruptly at the transition temperature exhibiting anomalous behaviour in the vicinity of the phase transition temperature. Investigations on the temperature variation study of the specific volume and the ultrasonic velocity in liquid crystals enables one to study and establish the transition temperatures as these will change abruptly at the transition. By using the specific volume and ultrasonic velocity data one can derive some more parameters like adiabatic compressibility (β_{ad}), molar sound velocity or Rao number (R_n), molar compressibility (B), van der Waals' constant (b) and also the intermolecular free lengths (L) as a function of temperature. In addition to these important parameters, the orientational order parameter (S) and its variation with temperature can also be studied. Even though such studies on liquid crystals are plenty in the literature (Natale; Rao, Murty; Choudary et al.), they are mainly limited to only to a single ultrasonic frequency in many cases.

The present studies deal with the estimation of the above mentioned parameters in the liquid crystal 4'-n-pentyloxy-4-cyanobiphenyl (5 OCB) using the specific volume and ultrasonic velocity data at four different fixed ultrasonic frequencies which forms part of our systematic studies on the liquid crystalline phase transitions (Rao, Murty; Choudary et al.; Rao et al., Venkatacharyulu et al.). The variation of the order parameter with temperature at the four different frequencies is also studied.

Experimental

The structural formula and the scheme of the transition temperatures measured in the present studies using a polarizing microscope with a hot stage arrangement in the liquid

crystal are

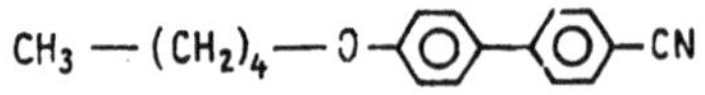

$$\text{Solid} \underset{320.5\,K}{\overset{320.5\,K}{\rightleftarrows}} \text{Nematic} \underset{340.5\,K}{\overset{340.5\,K}{\rightleftarrows}} \text{Isotropic}$$

4′-n-pentyloxy-4-cyanobiphenyl (5 OCB) was secured from M/s Holfmann-La-Roche, Switzerland, under the trade number liquid crystal component 5315. The phase transition temperatures are presented above along with the company quoted transition temperatures (indicated in brackets). As the compound was obtained in its pure form and as the measured transition temperatures are in agreement with the values given by the company, the compound was subjected to experimental investigations without any further purification.

A bicapillary pyknometer with a bore diameter of 0.2 mm is used in the present investigations to measure the specific volume at different temperatures. The variation of the ultrasonic velocity as a function of the temperature is studied at the four different fixed frequencies 0.8 MHz, 2 MHz, 3 MHz and 6 MHz, using the multifrequency ultrasonic interferometer supplied by Mittal Enterprises, New Delhi, India. The cell is essentially the same as that supplied with the interferometer except with a few modifications. The details regarding the experimental set up were described in detail elsewhere (VENKATACHARYULU et al.). The estimated accuracies in the measurements of specific volume, ultrasonic velocity and temperature are ± 0.0001 cc/gm, $\pm 0.2\%$ and ± 0.1 K, respectively.

Results and discussion

The variation of the specific volume with temperature in the liquid crystal 5315 which was studied in the temperature range 333–351 K is indicated in Figure 1. From the figure it can be readily observed that the specific volume decreases linearly with the decrease of temperature both in the isotropic phase as well as in the nematic phase except in the vicinity of isotropic-nematic transition where there is a sudden fall. The thermal expansion coefficients in the isotropic, nematic phases are found to be 8.21×10^{-4}/K at 351 K, 9.3×10^{-4}/K at 334 K, respectively and its value at the transition temperature 340.5 K is found to be 31×10^{-4}/K. The pretransitional effects are found to be predominant on the lower side, i.e., nematic side of the tran-

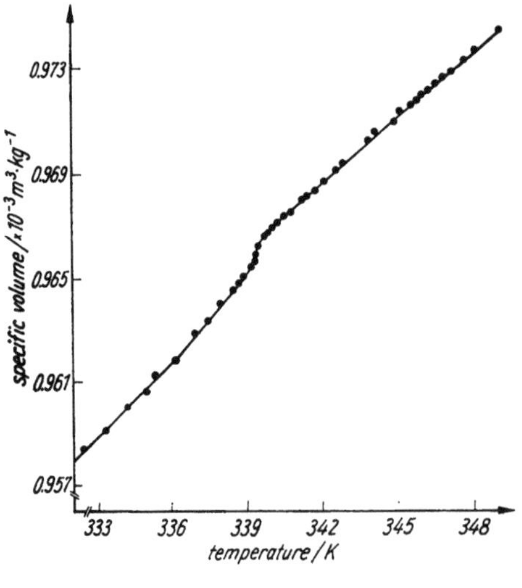

Fig. 1. Variation of specific volume with temperature in the liquid crystal 4′-n-Pentyloxy-4-cyanobiphenyl

sition. The specific volume jump at the transition temperature is found to be 0.12%. The high value of thermal expansion coefficient and the large change in the specific volume at the transition indicates that the transition can be of first order. The sudden decrease in specific volume near isotropic-nematic transition temperature is attributed to a sudden change in structure from a disordered and less densely packed liquid state to a more ordered and closely packed nematic phase. The thermal expansion coefficient in the nematic phase is observed to be larger when compared to its value in the isotropic state which is an indication of increasing order with decreasing temperature (BAHADUR; RAO et al.).

The variation of the ultrasonic velocity with temperature in this liquid crystal is studied in the temperature range 332–349 K at the four fixed frequencies which is represented in Figure 2. Using the specific volume (v) and the ultrasonic velocity (V) data, the adiabatic compressibility (β_{ad}), molar sound velocity or RAO number (R_n), molar compressibility (B), van der Waals constant (b) and the intermolecular free lengths (L) are estimated using the relations (RAO; JACOBSON)

$$\beta_{ad} = v/V^2 , \qquad R_n = Mv/(V)^{1/3} ,$$

$$B = Mv/(\beta_{ad})^{1/7} , \qquad b = Mv\left[1 - \frac{RT}{MV^2}\left\{\sqrt{1 + \frac{MV^2}{3RT}} - 1\right\}\right],$$

$$L = \left(\frac{K}{\beta_{ad}}\right)^{-1/\beta},$$

where M is the molecular weight, k and p are constants, however, they are to some extent temperature dependent (JACOBSON).

These studies reveal a single transition temperature in the range investigated at 340.5 K at all the four different frequencies. The transition temperature observed in the ultrasonic studies is in agreement with the value obtained from the specific volume studies. Figure 2 reveals that the ultrasonic velocity decreases with the increase of temperature in the isotropic phase as is the case for ordinary liquids. This can be due to the decrease of temperature the mean distance between the molecules decreases, leading to an increase in the potential energy of interaction which further leads to

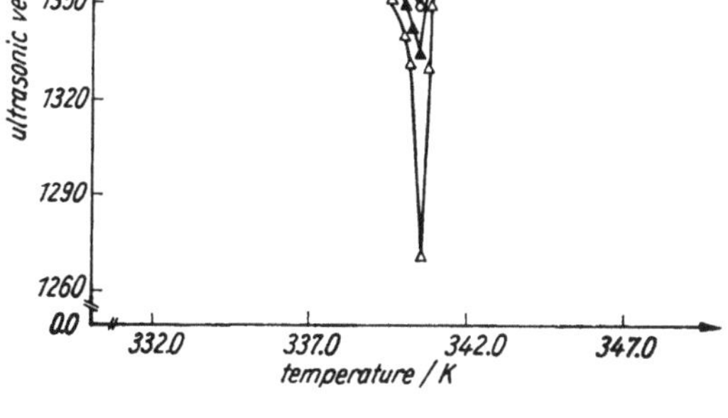

Fig. 2. Variation of ultrasonic velocity with temperature in the liquid crystal 4′-n-Pentyloxy-4-cyanobiphenyl. △ 0.8 MHz, ▲ 2 MHz, ○ 3 MHz, ● 6 MHz

an increase in the ultrasonic velocity. Just above the transition temperature the ultrasonic velocity decreases with the decrease of temperature attaining a minimum value at the isotropic-nematic transition and a further decrease in temperature results again to a rapid increase in the ultrasonic velocity in the mesophase at the four different frequencies and then increases linearly with the decrease of temperature. The entire course of this anomalous behaviour takes place within 3 K.

It is well-known that the ultrasonic velocity is related to the molecular structure, intensity and character of molecular interactions (BEYER, LETCHER; CANDAU, LETCHER). The sudden decrease in the velocity in the vicinity of isotropic-nematic transition can be attributed to the sudden structural changes that arise from the disordered isotropic liquid phase to the nematic phase which is associated with the long range orientational order of the molecules. The pretransitional effects occur on both sides of the isotropic-nematic transition temperature at all the four frequencies which can be explained on the basis of Frenkel's heterophase fluctuation theory (FRENKEL). The pre-transitional effects are found to be predominant on the lower side of the transition i.e., nematic side and similar behaviour was reported in many cases by earlier investigators (RAO, MURTY; BAHADUR). The pre-transitional effects on the higher side of the transition i.e., isotropic side can be attributed to the small heat of transition from the nematic phase to the isotropic liquid state.

The variation of the adiabatic compressibility (β_{ad}), with temperature for this liquid crystal at the various frequencies is indicated in Figure 3. Just like the ultrasonic velocity, the adiabatic compressibility also exhibits pre-transitional effects on either side of the transition, but they are predominant on the lower side, i.e., nematic side of the transition and the pre-transitional effects decrease with the increase of frequency. The anomalous effects increase with the decrease of frequency. The value of the adiabatic compressibility is less in the nematic phase when compared to its value in the isotropic liquid state for a particular frequency. But in the specific volume measurements the pre-transitional effects are observed mainly on the lower side of the transition as was observed in many cases by earlier workers (RAO, MURTY; RAO et al.).

The molar sound velocity or Rao number (R_n) and the molar compressibility (B) as a function of temperature are derived at the various frequencies and they are indicated in Table 1. These parameters exhibit interesting results. The variation of RAO number with temperature is found to be almost constant in each state with a high value in the isotropic state except in the vicinity of the isotropic-nematic transition. Just like the ultrasonic velocity the RAO number also increases with increasing fre-

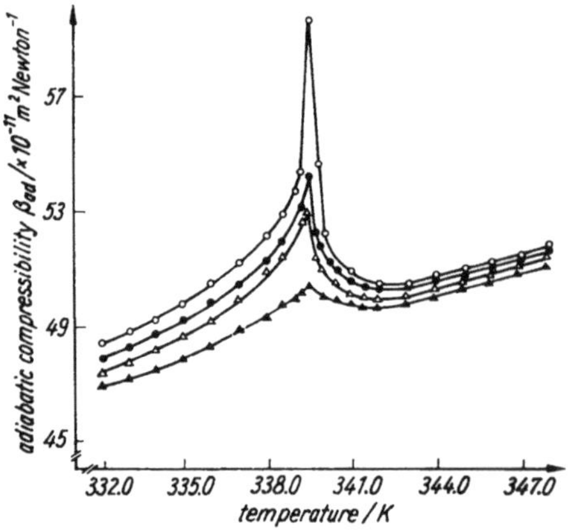

Fig. 3. Variation of adiabatic compressibility with temperature in the liquid crystal 4′-n-Pentyloxy-4-cyanobiphenyl. ○ 0.8 MHz, ● 2 MHz, △ 3 MHz, ▲ 6 MHz

Table 1
Molar Sound Velocity (R_n) and Molar Compressibility (B) at Different Temperatures of 4'-n-Pentyloxy-4-cyanobiphenyl

Sl No.	temperature K	0.8 MHz		2 MHz		3 MHz		6 MHz	
		R_n	B	R_n	B	R_n	B	R_n	B
1.	349.0	2871	7616	2873	7621	2874	7624	2877	7629
2.	348.0	2871	7516	2873	7621	2874	7624	2877	7629
3.	347.0	2870	7615	2872	7619	2874	7623	2876	7627
4.	346.0	2870	7613	2872	7619	2874	7623	2876	7627
5.	345.0	2869	7611	2872	7617	2873	7621	2875	7625
6.	344.0	2868	7610	2870	7614	2872	7618	2874	7624
7.	343.0	2865	7603	2867	7607	2870	7614	2873	7622
8.	342.5	2863	7599	2865	7594	2867	7608	2871	7617
9.	342.0	2859	7588	2861	7596	2865	7602	2869	7611
10.	341.5	2852	7573	2857	7584	2860	7591	2866	7604
11.	341.3	2850	7568	2853	7576	2858	7585	2864	7601
12.	341.0	2843	7553	2847	7562	2854	7576	2863	7597
13.	340.8	2821	7503	2842	7550	2850	7568	2861	7593
14.	340.5	2779	7406	2823	7507	2834	7531	2858	7589
15.	340.3	2819	7499	2827	7515	2835	7534	2858	7586
16.	340.0	2824	7509	2830	7524	2836	7537	2858	7588
17.	339.5	2829	7521	2837	7540	2842	7550	2859	7588
18.	339.0	2834	7532	2841	7548	2845	7558	2859	7589
19.	338.0	2839	7532	2846	7558	2850	7569	2861	7592
20.	337.0	2840	7546	2848	7563	2853	7574	2861	7592
21.	336.0	2843	7554	2849	7567	2855	7579	2861	7594
22.	335.0	2846	7561	2851	7572	2856	7581	2862	7597
23.	334.0	2847	7563	2852	7574	2857	7584	2863	7598
24.	333.0	2848	7564	2853	7575	2859	7587	2863	7598

R_n expressed in units of [cm^3 (m/sec)$^{1/3}$]; B expressed in units of [cm^3 (dyne/cm^2)$^{1/7}$]

quency both in the isotropic and nematic phases at any particular temperature. The behaviour of Rao number with temperature is similar to the ultrasonic velocity. Just as in the case of ultrasonic velocity the pre-transitional effects are predominant on the lower side of the transition, i.e., nematic side and found to be increasing with decreasing frequency. The behaviour of the molar compressibility is exactly similar to the Rao number as reported by the earlier workers over a number of thermotropic liquid crystals (Bahadur; Rao, Murty; Choudary et al.).

The van der Waals constant is calculated using the formula which was already given. The values of b for this liquid crystal in the nematic phase at a temperature of 334 K at 0.8 MHz, 2.0 MHz, 3.0 MHz, and 6 MHz are 245.00, 245.10, 245.40 and 245.30 whereas its value in the isotropic state at 348 K is 248.2, 248.2, 248.3 and 248.4. This indicates that the van der Waals' constant is independent on frequency and is constant for that phase. However, the value of b in the isotropic state is slightly greater than its value in the nematic phase. This shows that the actual volume of the molecules in a gram mole is constant as the van der Waals' constant represents four times the volume of the molecules in one gram mole.

The intermolecular free lengths (L) for this liquid crystal are calculated in the temperature range 349–333 K and they are represented in Table 2. The intermolecular free lengths decrease with the decrease of temperature both in the isotropic and mesophase states except at the transition. From Table 2 it can be seen, just like

Table 2
Intermolecular Free Lengths ($L \times 10^{11}$ m) 4'-n-Pentyloxy-4-cyanobiphenyl (5315)

Sl No.	temperature K	0.8 MHz	2 MHz	3 MHz	6 MHz
1	349.0	5.242	5.232	5.226	5.216
2	347.0	5.153	5.143	5.137	5.127
3	345.0	5.109	5.097	5.090	5.081
4	342.5	5.150	5.140	5.127	5.108
5	341.0	5.181	5.161	5.132	5.090
6	340.5	5.400	5.228	5.178	5.064
7	339.5	5.140	5.136	5.100	5.036
8	337.0	4.960	4.922	4.910	4.874
9	335.0	4.884	4.865	4.846	4.816
10	333.0	4.820	4.807	4.783	4.762

the ultrasonic velocity, adiabatic compressibility, etc., the intermolecular free lengths also exhibits anomalous behaviour in the vicinity of the transition temperature having a high value at the transition. The intermolecular free lengths decrease with the increase of the ultrasonic frequency at any particular temperature at which they are measured.

The order parameter at the transition temperature is estimated using the specific volume data adopting Maier-Saupe tables (Maier-Saupe) and the variation of the order parameter with temperature is estimated adopting the method suggested by (Chandrasekhar, Madhusudhana) at the four different frequencies and is represented in Figure 4. It can be observed from the figure that the order parameter increases with the decrease of temperature indicating an increasing order with the decrease of temperature. Further the order parameter decreases with the increase of frequency.

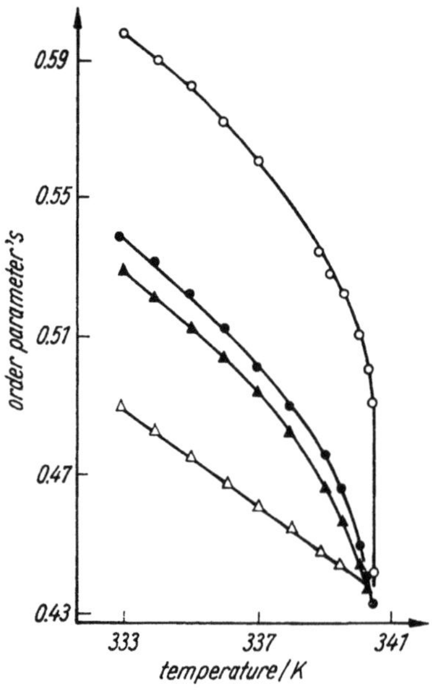

Fig. 4. Variation of order parameter with temperature in the liquid crystal 4'-n-Pentyloxy-4-cyanobiphenyl. ○ 0.8 MHz, ● 2 MHz, ▲ 3 MHz, △ 6 MHz

The decrease in the order parameter with the increase of frequency may be due to the fact, as the frequency increases the disturbance in the molecules also increases which leads to a decrease in the order parameter.

One of the authors (AVNG) thanks the authorities of T.J.P.S. College, Guntur, for allowing him to do part-time research.

References

BAHADUR, B.: Acoustica **34**, 486 (1975)
BAHADUR, B.: Acoustica **33**, 277 (1975)
BAHADUR, B.: J. Phys. (c), Solid State Physics, **9**, 11 (1976)
BEYER, R. T., LETCHER, S. V.: Physical ultrasonics, New York 1969
CANDAU, S., LETCHER, S. V.: Advances in Liquid Crystals **3**, 168 (1978)
CHANDRASEKHAR, S., MADHUSUDHANA, N. V.: Liq. Crystals — Part II, edited by G. H. BROWN, New York 1969, p. 291
CHOUDARY, L. V., RAO, J. V., MURTY, P. N., MURTY, C. R. K.: phys. stat. sol. **74**, 431 (1982)
FRENKEL, J.: Kinetic Theory of Liquids, New York 1946
JACOBSON, B.: Acta Chemica Scandinavica **6**, 1485 (1952)
JACOBSON, B.: J. Chem. Phys. **20**, 927 (1952)
MAIER, W., SAUPE, A.: Z. Naturforsch. **159**, 287 (1960)
NATALE, G. G.: J. Acoust. Soc. Amer. **65**, 1265 (1978)
RAO, J. V., RAO, K. R. K., CHOUDARY, L. V., VENKATACHARYULU, P., BALIAH, V.: Phase Transitions (in press)
RAO, J. V., MURTY, C. R. K.: Z. Naturforsch. **36a**, 1002 (1981)
RAO, J. V., RAO, K. R. K., CHOUDARY, L. V., MURTY, C. R. K.: Phase Transitions **5**, 73 (1986)
RAO, K. R. K., RAO, J. V., VENKATACHARYULU, P., BALIAH, V.: Mol. Cryst. Liq. Cryst. **136**, 307 (1986)
RAO, K. R. K., RAO, J. V., CHOUDARY, L. V., VENKATACHARYULU, P., BALIAH, V.: Z. Phys. Chem. (in Press)
RAO, K. R. K., RAO, J. V., CHOUDARY, L. V., MURTY, C. R. K.: Acta Physica Polonica **A63**, 419 (1983)
RAO, M. R.: Ind. J. Phys. **14**, 109 (1940)
RAO, M. R.: J. Chem. Phys. **9**, 682 (1941)
VENKATACHARYULU, P., GUPTA, A. V. N., RAO, J. V.: Cryst. Res. Technol. **22**, **743** (1987)

(Received February 13, 1987)

Authors' address:

Dr. A. V. N. GUPTA, J. V. RAO, P. VENKATACHARYULU
Department of Physics, Nagarjuna University
Nagarjunanagar 522510, India

Cryst. Res. Technol.	22	1987	10	1320

Book Review

G. E. BACON (ed.)

Fifty years of neutron diffraction. The advent of neutron scattering

Adam Hilger, Bristol 1986, xii and 280 pp., 92 figures, 7 tables, 30 £. ISBN 0-85274-587-7

R. J. WEISS writes in this book: "... that scientists make poor historians. They tend to focus in their hindsight on their own scientific achievements and less on matters of interest of historians (who are after the bigger picture, i.e. how do the interactions of poeple and ideas turn into history)". The paper by R. J. WEISS himself does not justify such an opinion concerning neutron diffraction. Moreover, the most famous expert in neutron diffraction, G. E. BACON, who has edited this excellent and instructive book, has also given both a superb historical preface and at the same time a such introduction to the subject.

Let me quote R. J. WEISS once more: "CYRIL SMITH sat on a committee in 1946 (with FERMI) which decided that neutron diffraction was not worth support since X-rays could perform the same function. We now may laugh at this." Why do we laugh ? The answer is given both by the 50th jubilee and by the headlines of the chapters without mentioning the exquisite pictures in the book: 1) Introduction — the pattern of 50 years, 2) The first diffraction experiments: the three papers of 1936, 3) Looking back: early days and memories, 4) World-wide spread of neutron scattering a miscellany of stories, 5) The changing scene: the appearence of high-flux reactor, 6) Some statistics from neutron diffraction commission, 7) Some crucial applications, 8) Changing techniques, 9) Some views of the future: reactor and accelerator.

Only one example shall be given here: "Neutrons in biology", the article by B. JACROT. The application of neutron diffraction in biology is based on the complementarity of X-rays and neutrons: e.g. location of hydrogen atoms is possible. The first investigation was the analysis of vitamin B_{12} in 1965. In the following years further experiments were carried out with proteins, DNA, membranes, viruses and cells. Many biological structures and their dynamics could be revealed only by neutron diffraction. One might speak about biological neutron diffraction "crystallography". We can conclude that not only physics and history are of great importance with regard to neutron diffraction but also biology and history.

D. LEUSCHNER

Cryst. Res. Technol.	22	1987	10	1321–1331

G. Pelzl, U. Böttger, S. Diele, D. Demus

Forschungsabteilung der Sektion Chemie, Martin-Luther-Universität Halle-Wittenberg, DDR

Induced S_A Phases in Binary Systems with EDA Complexing

The phase diagrams of 13 binary systems are presented in which electron donor-acceptor (EDA) complexing is clearly detectable. In all binary systems studied the same electron donor compound was used whereas the electron acceptor compounds were members of two homologous series. It was found that the induction of smectic A phases increases with increasing length of the alkyl chain of the acceptor compound and with increasing stability of the complexes.

Die Phasendiagramme von 13 binären Systemen werden vorgestellt, für welche EDA-Komplexbildung deutlich nachweisbar ist. In allen untersuchten binären Systemen wurde die gleiche Elektronendonator-Verbindung eingesetzt, während als Elektronenakzeptor-Verbindungen Glieder von zwei homologen Reihen dienten. Es wurde gefunden, daß die Induktion von S_A-Phasen mit wachsender Alkylkettenlänge der Akzeptorverbindung sowie mit erhöhter Stabilität der Komplexe zunimmt.

1. Introduction

In some binary systems of liquid crystals smectic A (S_A) phases may occur at middle concentrations with a maximum in the transition curve S_A/nematic or S_A/isotropic, respectively, even if the pure compounds do not exhibit any smectic phase (Dave et al.; Schroeder, Schroeder).

Such smectic phases are generally called "induced" smectic phases. The term "induced" is used, too, if in binary systems of smectogenic compounds the temperature of the existence region of a smectic phase is markedly enhanced (Sackmann, Demus). The origin of the induced smectic phases is not likely to be uniform. But there are experimental evidences that in some cases electron-donor-acceptor (EDA) interaction has a stabilizing effect on the smectic phases (Sharma et al.; Demus et al.; Schneider, Sharma; Araya, Matsunaga). In this paper we present new examples of induced S_A phases due to EDA complex formation. In all binary systems studied the same electron-donor compound was used which only exhibits a nematic phase. This compound was combined with members of two homologous series which possess electron-acceptor properties. In this way the influence of the length of the alkyl chain of the acceptor compound on the S_A induction should be investigated.

2. Materials

1. Electron-donor compound:

4,4′-bis-[n-butylamino]-biphenyl

(Sharma, Weissflog et al.)

K 73 N 96 Is

2. Electron-acceptor compounds:

a) 4-n-alkyloxyphenyl 4-nitrocinnamates

(KUSCHEL, DEMUS)

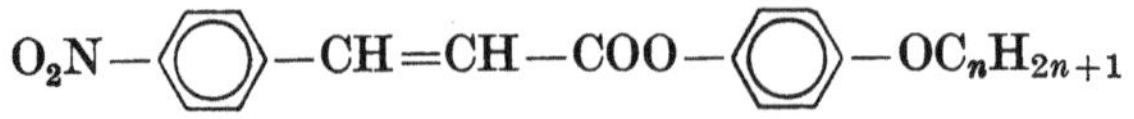

n	K	S_{Ad}[1])	N	Is
4	. 128	— —	(. 124)	.
5	. 99	— —	. 123	.
6	. 109	— —	. 127	.
7	. 114	— —	. 129	.
8	. 104	. 112	. 130	.

b) 4-β-cyanethylphenyl 4-n-alkyloxycinnamates

(ZASCHKE et al.)

n	K	S_{Cd}[1])	S_{Ad}[1])	N	Is
3	. 87.2	— —	— —	. 110	.
5	. 91	— —	(. 31[2])	. 106.5	.
6	. 71.5	— —	— —	. 109.2	.
7	. 70	— —	— —	. 106	.
8	. 87	— —	— —	. 108.2	.
9	. 65	(. 53)	. 105.2	. 107	.
10	. 69	(. 60)	. 110	. 110.5	.
12	. 82	(. 58)	. 117	— —	.

K: solid crystal Is: isotropic liquid
N: nematic
S_{Ad}, S_{Cd}: smectic A_d ... C_d

The numbers between the phase symbols are the transition temperatures (°C) Brackets denote monotropic phases.

3. Experimental

The phase diagrams were studied with a polarizing microscope by means of the contact method (SACKMANN, DEMUS) and by the determination of the transition temperatures in mixtures of singular concentrations.

The layer spacings of the smectic phases were measured by X-ray investigations using a small angle equipment.

4. Phase diagrams

4.1. Binary systems with homologous 4-n-alkyloxyphenyl 4-nitrocinnamates

In all binary systems of the nitrocinnamates with 4,4′-bis-[n-butylamino]-biphenyl the mixed phases of the isotropic and liquid crystalline state show a strong red-

[1]) According to X-ray investigations the S_A as well the S_C phases are of the bilayer type (S_{Ad} resp. S_{Cd}).

[2]) This metastable S_A phase probably is of the monolayer type.

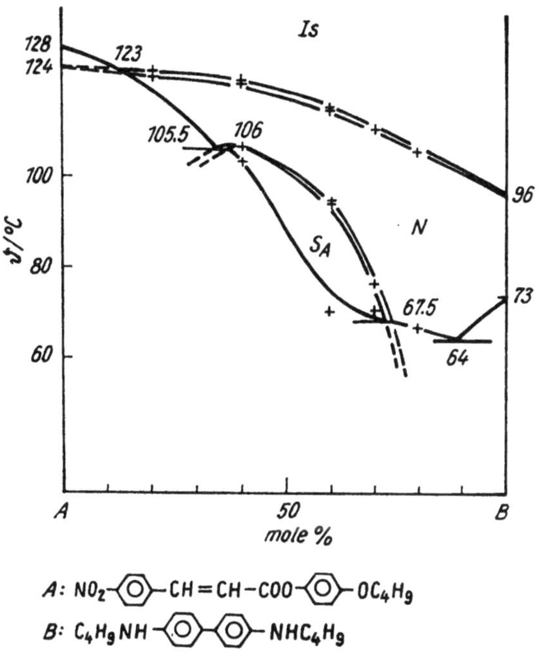
Is
128
124
123
105.5
106
100
ϑ/°C
80
60
S_A
N
96
73
67.5
64
A
50
mole %
B
A: NO2-⟨O⟩-CH=CH-COO-⟨O⟩-OC4H9
B: C4H9NH-⟨O⟩-⟨O⟩-NHC4H9

Fig. 1

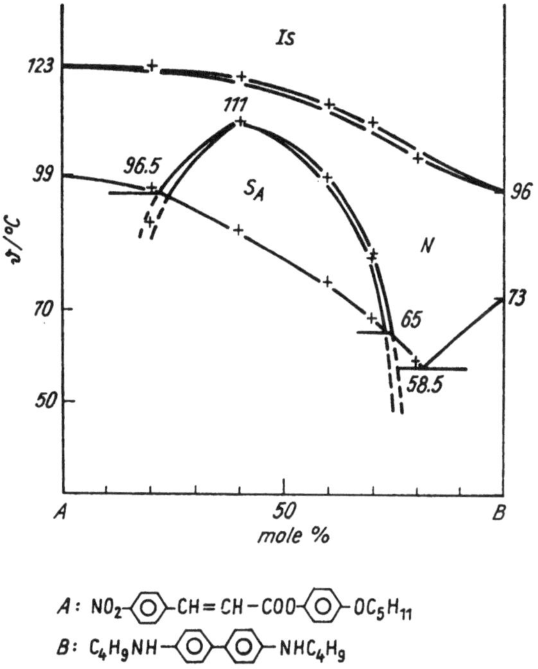
Is
123
111
96.5
99
S_A
ϑ/°C
N
96
70
73
65
58.5
50
A
50
mole %
B
A: NO2-⟨O⟩-CH=CH-COO-⟨O⟩-OC5H11
B: C4H9NH-⟨O⟩-⟨O⟩-NHC4H9

Fig. 2

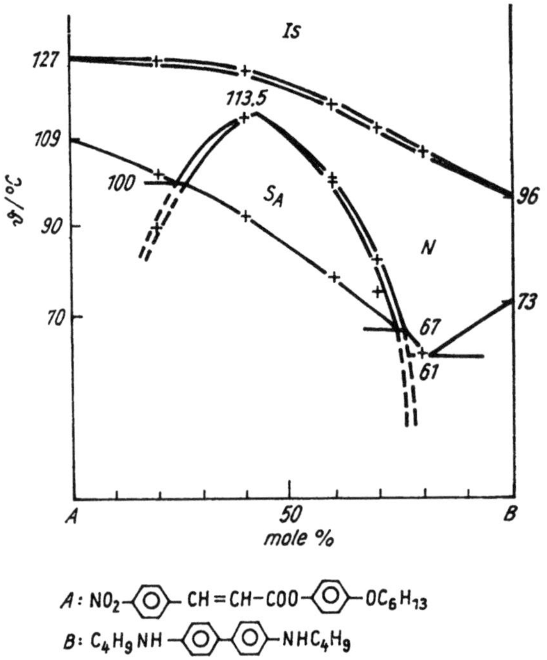
Is
127
113.5
109
100
S_A
ϑ/°C
90
N
96
70
73
67
61
A
50
B
mole %
A: NO_2–⟨◯⟩–CH=CH–COO–⟨◯⟩–OC_6H_{13}
B: C_4H_9NH–⟨◯⟩–⟨◯⟩–NHC_4H_9

Fig. 3

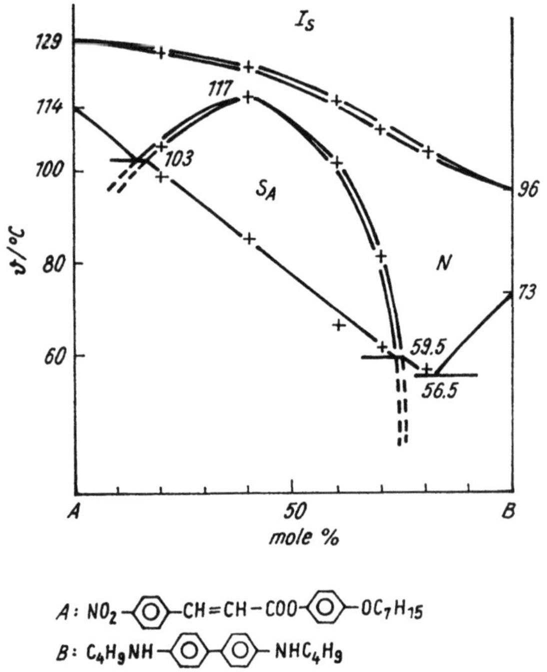
I_S
129
117
114
103
100
S_A
96
ϑ/°C
80
N
73
60
59.5
56.5
A
50
B
mole %
A: NO_2–⟨◯⟩–CH=CH–COO–⟨◯⟩–OC_7H_{15}
B: C_4H_9NH–⟨◯⟩–⟨◯⟩–NHC_4H_9

Fig. 4

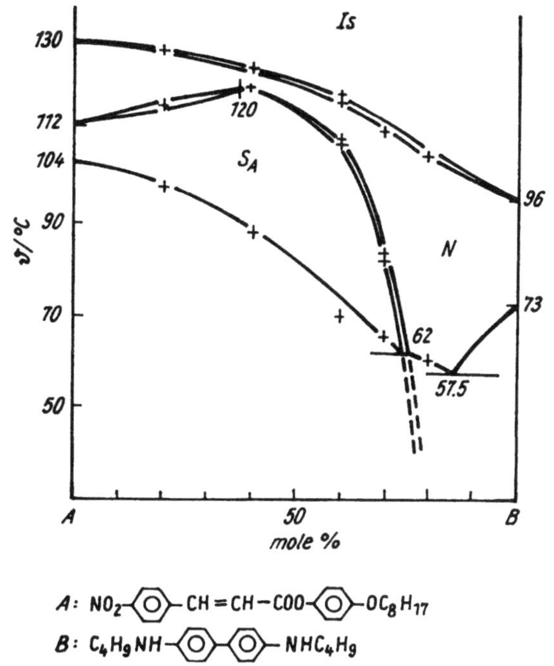

Fig. 5

brown colour whereas the single components are nearly colourless. This colour is due to a charge-transfer (CT) band between 475 and 485 nm (SHARMA et al.).

The phase diagrams are shown in the Figures 1—5. The common feature of these diagrams is the occurrence of an induced S_A phase with the characteristic maximum of the transition curve S_A/N. The maximum in all cases lies at about 60 mole% of component A. The temperature of the maximum continuously increases with increasing length of the alkyl chain of the acceptor compound A. It is obvious that the complexing also enhances the transition temperatures N/Is.

According to X-ray investigations the S_A-phase of 4-n-octyloxyphenyl-4-nitrocinnemates is of the S_{A_d} type because the layer spacing d is 34.1 Å, but the molecule length L is only 27.7 Å ($d/L = 1.23$). With increasing content of the terminal-nonpolar compound B the ratio $d/\bar{L}$ decreases, where $\bar{L}$ is the average molecule length $\bar{L} = x_A L_A + x_B L_B$ (x_A, x_B: molar fractions of the components A and B). For example, in a mixture of 60 mole % 4-n-octyloxyphenyl 4-nitrocinnamate (in the region of the induced S_A phase, see Fig. 5) the ratio $d/\bar{L}$ is nearly 1 ($d = 26$ Å; $\bar{L} = 25.6$ Å) indicating a monolayer S_A phase.

4.2. Binary systems with 4-β-cyanethylphenyl-4-n-alkyloxycinnamates

The colourless 4-β-cyanethylphenyl 4-n-alkyloxycinnamates give with the colourless 4,4′-bis-[butylamino]-biphenyl yellow mixed phases in the isotropic and liquid crystalline state (CT band at about 400 nm). In the binary systems of the methoxy, ethoxy, propyloxy and butyloxy homologues no S_A phases occur as seen for the example of the propyloxy homologue (see Fig. 6). But in all phase diagrams of this type the transition curve shows a distinct minimum.

In the diagrams of the higher homologues (C_5—C_{12}) metastable induced S_A phases occur (see Figs. 7—13). For the pentyloxy homologue the induced S_A phase is starting

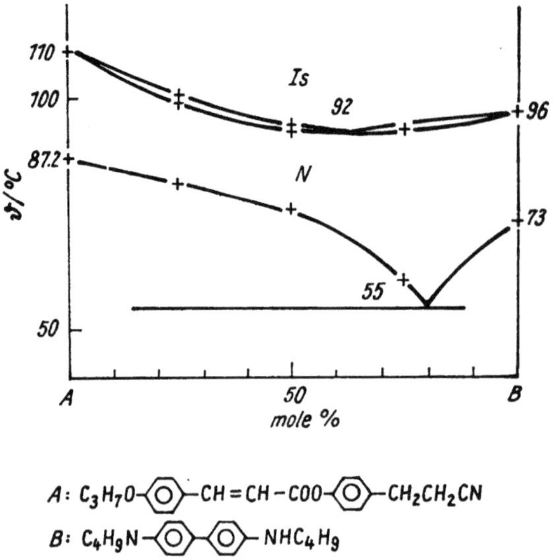

Fig. 6

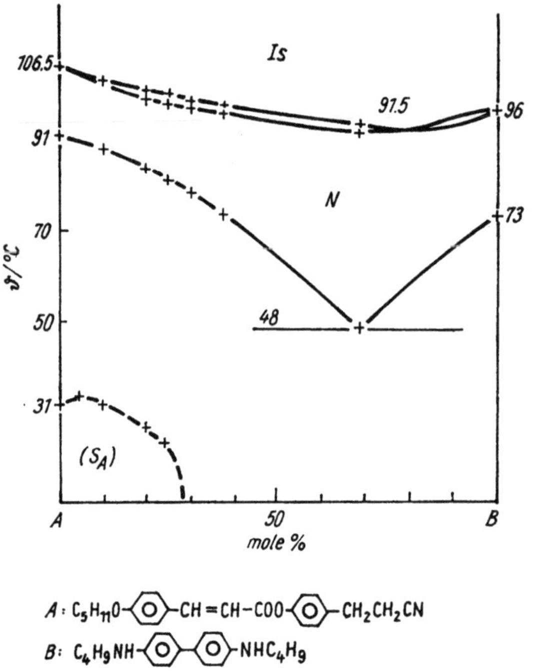

Fig. 7

from a metastable S_A phase of the pure component A (Fig. 7). The maximum of the transition curve S_A/N is at about 90 mole% of component A. For the homologues C_6 to C_8 (also for the C_4 homologue) a monotropic S_A phase was not detectable, also not by rapid supercooling of the nematic phase to 0 °C. The maximum of the transition curve S_A/N lies at about 80 mole% of the acceptor compound A (see Figs. 8—10). But this maximum shows an increasing tendency from the C_5 homologue to the C_8 homologue (Figs. 7—10).

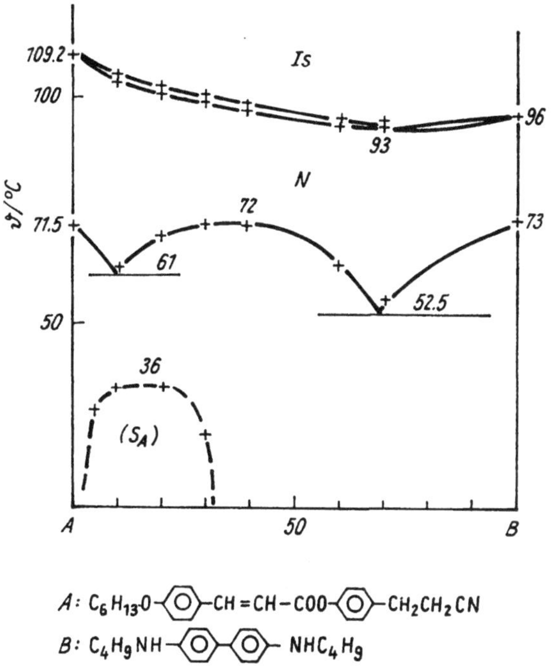
Is
109.2
100
96
93
N
72
71.5
73
61
52.5
50
36
(S_A)
ϑ/°C
A
50
B
A: $C_6H_{13}O$ CH=CH-COO CH_2CH_2CN
B: C_4H_9NH NHC_4H_9

Fig. 8

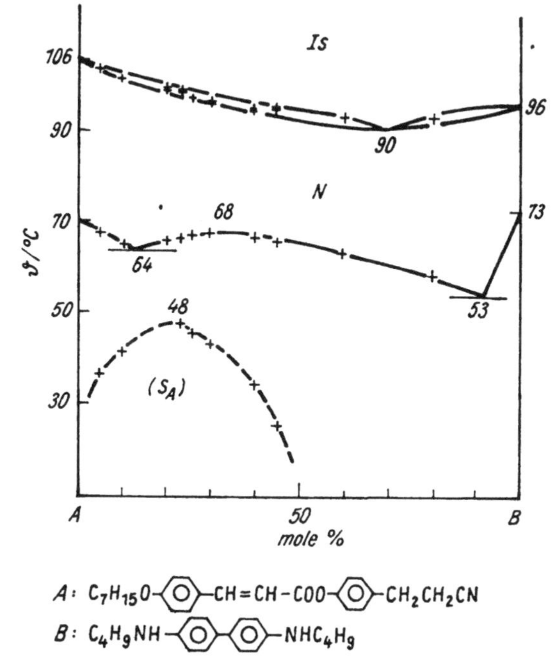
Is
106
90
96
90
N
68
70
73
64
53
50
48
(S_A)
30
ϑ/°C
A
50
mole %
B
A: $C_7H_{15}O$ CH=CH-COO CH_2CH_2CN
B: C_4H_9NH NHC_4H_9

Fig. 9

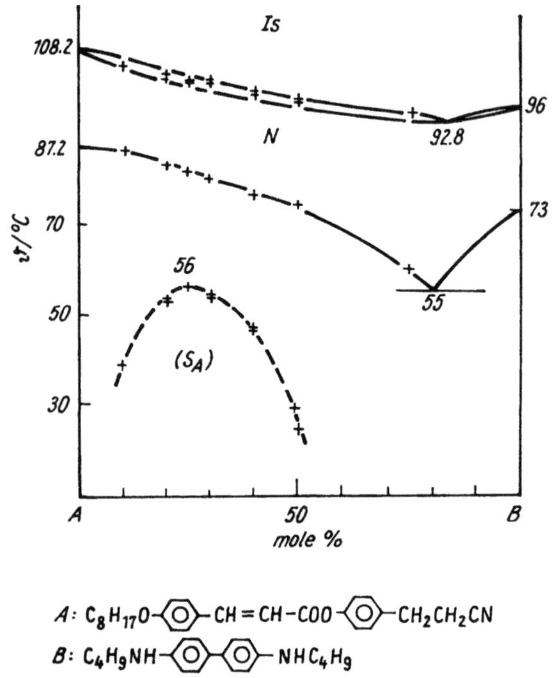
Is
108.2
96
N
92.8
87.2
73
ϑ/°C
70
56
55
50
(S_A)
30
A
50
B
mole %
A: $C_8H_{17}O$-⟨○⟩-CH=CH-COO-⟨○⟩-CH_2CH_2CN
B: C_4H_9NH-⟨○⟩-⟨○⟩-NHC_4H_9

Fig. 10

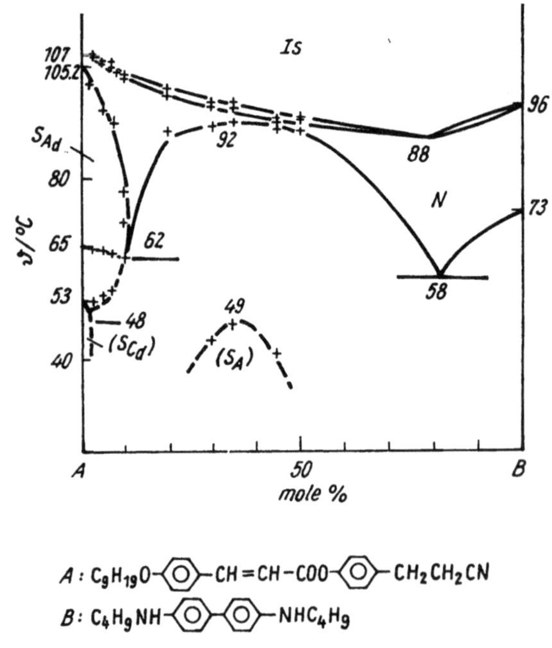
Is
107
105.2
96
S_{Ad}
92
88
80
N
73
ϑ/°C
65
62
53
58
49
48
(S_{Cd})
(S_A)
40
A
50
B
mole %
A: $C_9H_{19}O$-⟨○⟩-CH=CH-COO-⟨○⟩-CH_2CH_2CN
B: C_4H_9NH-⟨○⟩-⟨○⟩-NHC_4H_9

Fig. 11

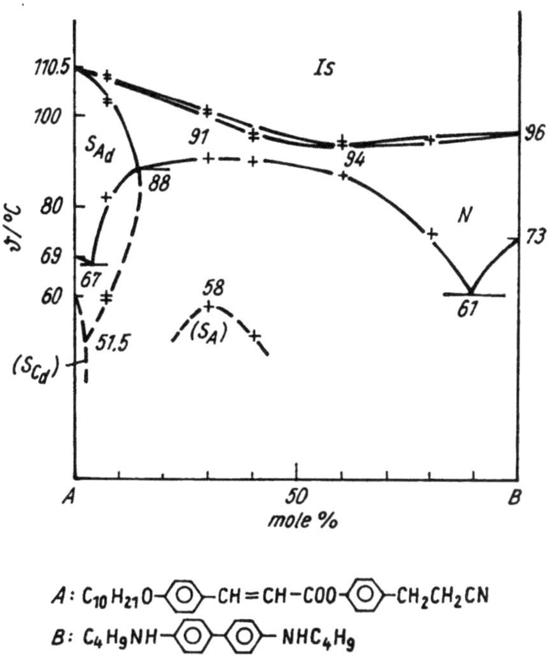
Is
110.5
100
80
69
60
ϑ/°C
S_{A_d}
91
94
96
88
N
73
67
58
61
51.5
(S_A)
(S_{C_d})
A
50
B
mole %
A: $C_{10}H_{21}O$–⟨O⟩–CH=CH–COO–⟨O⟩–CH_2CH_2CN
B: C_4H_9NH–⟨O⟩–⟨O⟩–NHC_4H_9

Fig. 12

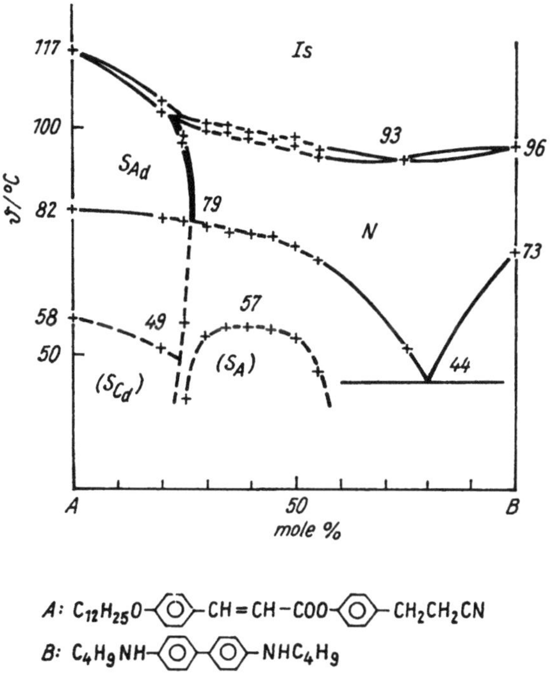
Is
117
100
82
58
50
ϑ/°C
S_{A_d}
93
96
79
N
73
57
49
44
(S_A)
(S_{C_d})
A
50
B
mole %
A: $C_{12}H_{25}O$–⟨O⟩–CH=CH–COO–⟨O⟩–CH_2CH_2CN
B: C_4H_9NH–⟨O⟩–⟨O⟩–NHC_4H_9

Fig. 13

Much more complicated diagrams were found for the higher homologues (C_9, C_{10}, C_{12}) which form stable S_{Ad} phases. It is seen from Figs. 11—13 that the S_{Ad} phase region which is restricted to mixtures of higher concentrations of component A increases with increasing chain length of component A. On cooling of the S_{Ad} phase a monotropic reentrant nematic phase appears. Between the S_{Ad} and the reentrant nematic phase a S_C phase occurs in a limited concentration range. Similar to the diagrams of the lower homologues an induced S_A phase region is observed which is not linked with the S_A phase region. The maximum of the induced S_A phase is somewhat shifted to lower concentrations of component A. Altough the maximum in the diagram of the C_9 homologue is lower than in the diagram of the C_8 homologue, among the diagrams with S_{Ad} phases the maximum temperature S_A/N has an increasing tendency.

On the other hand, it should be emphasized that in some phase diagrams pronounced maximum of the melting curves are observed (Figs. 8, 9, 11, 12) indicating complex formation in the solid state.

5. Discussion

By use of homologous compounds we were able to study the enhancement of the S_A phase region in binary systems with EDA interaction in dependence of the alkyl chain length of the electron acceptor compound. As can be seen in the diagrams presented, generally the maximum temperature of the S_A/N transition curve increases with increasing length of the acceptor compound. A modified behaviour was found for those cyanethyl compounds which exhibit a S_{Ad} phase, too (Figs. 11—13). In this case the S_{Ad} region is separated from the region of the induced S_A phase by a nematic region. The concentration range of the S_{Ad} phase increases with increasing chain length whereas the region of the induced S_A phase (and their maximum) is shiftet to mixtures of lower concentrations of the acceptor compound. Also the maximum of the S_A/N transition curve is lowered if the acceptor compound exhibits a S_{Ad} phase (compare Figs. 10 and 11). It should be noted that the increase of S_A induction with increasing alkyl chain length of the acceptor compound was found also for other binary systems (DEMUS et al. 1981).

Unfortunately, the S_A phase type of 4-β-cyanethylphenyl 4-n-pentyloxycinnamate could not be identified by X-ray investigations because the S_A phase crystallized in a relatively short time. The absence of the S_A phase in the next homologues (C_6 to C_8) indicates that the S_A phase of the pentyloxycinnamate should be different in type compared with that of the nonyloxycinnamate. If the S_A phase of the pentyloxycinnamate would be a S_{A1} phase we must expect S_A–N transitions in the C_6, C_7 and C_8 members at higher temperatures than for the pentyloxycinnamate according to the general trend of this phase transition within homologous series (DEMUS, DEMUS, ZASCHKE). It is known that in homologous series of terminal-polar compounds the lower members of the series have a monolayer S_{A1} phase and the higher members have a S_{Ad} phase (HEPPKE et al.). In this case the transition temperatures S_{Ad}/N as function of the length of the alkyl chain shows a maximum. Considering these results it can be assumed that in the homologous 4-β-cyanethylphenyl 4-n-alkyloxycinnamates not only the pentyloxy homologue but also the neighbouring members of the series (C_4, C_6, C_7) can exhibit S_{A_1} phases which are not detectable because of an unsufficient supercooling of the nematic phase.

In the binary systems presented EDA complexing was clearly detected. In analogous series of complexes the wavelength of the CT band rises with increasing complex formation constant. Therefore, the EDA complexes of the cyanethyl compounds

($\lambda_{CT} \sim 400$ nm) should be less stable than those of the nitrocinnamates ($\lambda_{CT} \sim 480$ nm). By comparing the phase diagrams of both homologous series it is obvious that the stronger EDA interaction of the nitrocinnamates (Figs. 1—5) leads to stronger S_A induction.

Also in some series of binary systems where EDA complexing cannot be detected the S_A induction was studied in dependence on the molecule lengths of the components (SZABON; SZABON et al.). Similar to the systems presented here, an increase of S_A induction was found with increasing molecule length of one component. But if this component reaches a definite molecule length (if the ratio of the molecule lengths of the components has a critical value) the S_A induction shows a saturation. This result is reasonable because increasing differences of the molecule lengths obviously lead to a lowering of the smectic order. This effect counteracts to the general increase of smectogenic properties with increasing length of the alkyl chains within homologous series.

The question arises why in our systems such a saturation of smectic induction was not observed. This result can be due to the EDA interaction which causes a stronger lateral cohesion of the molecules so that such saturation of smectic induction probably is shiftet to greater differences of the molecule lengths of the mixing components.

References

ARAYA, K., MATSUNAGA, Y.: Mol. Cryst. Liq. Cryst. **67**, 153 (1981)

DAVE, J. S., PATEL, P. R., VASANTH, K. L.: Ind. J. Chem. **4**, 505 (1966)

DE JEU, W. H., LONGA, L., DEMUS, D.: J. Chem. Phys. **84**, 6410 (1986)

DEMUS, D., DEMUS, H., ZASCHKE, H.: Flüssige Kristalle in Tabellen, Leipzig 1974

DEMUS, D., PELZL, G., SHARMA, N. K., WEISSFLOG, W.: Mol. Cryst. Liq. Cryst. **76**, 241 (1981)

HEPPKE, G., HOPF, R., KOHNE, B., PRAEFCKE, K.: in: Advances in Liquid Crystal Research and Applications, ed. by L. BATA, Oxford, Budapest 1980, p. 141

KUSCHEL, F., DEMUS, D.: Z. Chem. **15**, 350 (1975)

SACKMANN, H., DEMUS, D.: Z. phys. Chem. (Leipzig) **222**, 143 (1963)

SCHNEIDER, F., SHARMA, N. K.: Z. Naturforsch. **36a**, 1086 (1981)

SCHROEDER, J. P., SCHROEDER, D. C.: J. Org. Chem. **33**, 591 (1968)

SHARMA, N. K., PELZL, G., DEMUS, D. WEISSFLOG, W.: Z. phys. Chem. (Leipzig) **261**, 579 (1980)

SHARMA, N. K., WEISSFLOG, W., RICHTER, L., DIELE, S., WALTER, B., SACKMANN, H., DEMUS, D.: Advances in Liquid Crystal Research and Applications, Oxford, Budapest, Vol. 1, p. 3 (1980)

SZABON, J.: Abstracts of the 10th Internat. Liquid Crystal Conference, E 12, York 1984

SZABON, J., WEISSFLOG, W., PELZL, G., DIELE, S., DEMUS, D.: Cryst. Res. Technol. **21**, 1097 (1986)

ZASCHKE, H., MEWES, R., KÖNIG, S., SKUBATZ, R., DEMUS, D.: Z. Chem. **17**, 370 (1977)

(Received March 9, 1987)

Authors' addresses:

Dr. G. PELZL, Dr. U. BÖTTGER, Dr. S. DIELE, Prof. Dr. habil. D. DEMUS
Forschungsabteilung der Sektion Chemie
Martin-Luther-Universität Halle-Wittenberg
Mühlpforte 1
Halle/Saale
DDR-4020

Book Review

R. M. ALMEIDA (ed.)

Halide Glasses for Infrared Fiberoptics

Proceedings of the NATO Advanced Research Workshop on "Halide Glasses for Infrared Fiberoptics", Vilamoura (Algarve), Portugal, March 31—April 4, 1986
NATO ASI series. Series E, Applied sciences; no. 123

Martinus Nijhoff Publishers, Dordrecht, 1987. US $ 195.00. ISBN 9024-734800

In diesem Band sind 30 Originalbeiträge von 35 Experten aus 14 Ländern vereinigt. Schwermetallhalogenidgläser (vor allem Fluoride) werden erst seit 10 Jahren synthetisiert und untersucht, jedoch erschienen schon 1985 über 800 Publikationen zu diesem Gebiet. Das enorme wissenschaftliche Interesse wird durch den potentiellen ultrageringen Transmissionsverlust des Zirkonfluoridglases im Wellenlängenbereich von 2—3 μm hervorgerufen, der bei etwa 10^{-2} bis 10^{-3} db/km erwartet wird. Außerdem zeichnet sich dieses Material durch eine gute thermische und chemische Stabilität aus. Es ist somit ein äußerst geeigneter Kandidat für weitreichende Kommunikationssysteme auf der Basis optischer Fasern (Lichtleiter).

Der gegenwärtige Stand auf dem Gebiete der Infrarot-Faseroptik wurde sehr kritisch und fundiert dargestellt. Das äußerst sich auch in der Gliederung der Beiträge in 1. Rchstoffe und Reinheit, 2. Struktur, 3. Thermische und mechanische Erscheinungen, 4. Optische Eigenschaften der Gläser und Fasern, 5. Faserfabrikation, 6. Fasersysteme und 7. Andere Halogenidgläser. Derzeitig ist es praktisch möglich, insbesondere durch reine Ausgangsmaterialien und entsprechende Schmelzbedingungen, optische Fasern (Zirkonfluoridglas) mit einem Transmissionsverlust von 1 db/km bei 2,55 μm herzustellen. Aussagen darüber, wie der zu erwartende äußerst geringe Transmissionsverlust zu erreichen ist, wurden nur sehr vorsichtig getroffen. Es wurde auch die Absicht dieser Konferenz, die Forschungsrichtungen der nächsten fünf Jahre auf dem Gebiet der Infrarot-Lichteliter abzustecken, nicht umfassend dargelegt. Jedoch zeichnet sich ab, daß für praktische Zwecke die VD-Methoden favorisiert sind. Die Anwendung anderer Halogenide, wie Chloride, Bromide und Jodide, scheint größere Schwierigkeiten, insbesondere wegen der Beständigkeit, zu bereiten.

Die hauptsächlichen Ursachen für den geringen Transmissionsverlust im 2—3 μm Bereich, der reduzierte Rayleigh-Streuverlust sowie die vernachlässigbare Multiphononabsorption sind durch die ungewöhnliche Nahordnungsstruktur des Zirkonfluoridglases bedingt. Durch Kombination und Vergleich der verschiedenen Artikel können gerade bezüglich des Zusammenhanges von Struktur, Eigenschaften und Herstellungsbedingungen verallgemeinernde Aussagen getroffen werden.

In diesem Sinne ist dieser Band nicht nur den Spezialisten auf dem Gebiete der Lichtleiter zu empfehlen sondern auch Physikern Chemikern und Werkstoffwissenschaftlern, welche sich mit der Entwicklung neuer Gläser und anderer Werkstoffe beschäftigen. Ebenso findet der meßmethodisch interessierte und orientierte Festkörperwissenschaftler moderne Darstellungen zur Spektroskopie, zu den Diffraktionsmethoden sowie zu den Messungen unterschiedlicher Glaseigenschaften. Der Wert der einzelnen Artikel hat auch dadurch sichtlich zugenommen, indem die Fragen und Diskussionen mit in den Band aufgenommen wurden. Insgesamt liegt sowohl inhaltlich als auch in der Darstellung ein gelungener zusammengfassender Überblick über ein sich wegen seiner großen praktischen Bedeutung schnell entwickelndes Gebiet der Physik, Chemie und Technologie der Halogenidgläser vor.

B. MÜLLER

Cryst. Res. Technol.	22	1987	10	1333–1339

G. HAUCK, H. D. KOSWIG

Zentralinstitut für Elektronenphysik, Akademie der Wissenschaften der DDR, Berlin

Helical Structure and Threshold Voltage of Cholesteric-Nematic Phase Change at Different Anchoring Conditions

The helical structure of cholesteric mixed systems and their electro-optical behaviour are studied at planar and pretilted surface orientation. The results indicate that the influence of the forces of wall orientation is decreased by the helical structure. To prove this model strong and weak anchoring conditions of the molecules in the cholesteric mixture at the substrate surface are realized. Also the behaviour of the electro-optical threshold voltage at different surface conditions supports the assumptions concerning the helical structure.

Die helikale Struktur von cholesterinischen Mischungssystemen und ihr elektrooptisches Verhalten wurden bei planarer und angekippter Oberflächenorientierung untersucht. Die Resultate weisen darauf hin, daß der Einfluß der Kräfte der Grenzflächenorientierung durch die helikale Struktur verringert ist. Um dieses Modell zu überprüfen, wurden starke und schwache Kopplungsbedingungen der Moleküle der cholesterinischen Mischung an der Substratoberfläche realisiert. Auch das Verhalten der elektro-optischen Schwellspannung bei unterschiedlichen Oberflächenbedingungen unterstützt die Annahmen, die die helikale Struktur betreffen.

1. Introduction

The field-induced cholesteric-nematic phase change is of great interest for constructing of liquid crystal (LC) display devices. They offer the possibility to work without polarizers and show a good angular distribution of contrast. The cases of planar (surface tilt angle $\varphi_s = 0°$) and homeotropic ($\varphi_s = 90°$) surface alignment of the LC molecules are well studied (LIN-HENDEL). The electro-optical behaviour of such systems is essentially defined by the ratio d/P_0, where d is the thickness of liquid crystal layer and P_0 is the natural pitch. In general, at arbitrary φ_s the situation is not sufficiently clarified (TOGASHI et al.). Only the bistable cells with pretilted surface orientation and small twist angles ($\alpha \leqq 2\pi$) are more extensively investigated (BERREMAN).

It is the purpose of this paper to take into consideration some aspects of the influence of different pretilt angles on the helical structure and the threshold voltage of the phase change. The electro-optical behaviour depends also on the molecular orientational distribution in the bulk of cells at zero field. Measurements of the optical path difference Γ and of the mean tilt angle $\overline{\varphi}$ were performed to prove a model of internal structure of the twisted orientational pattern. Weak anchoring between the LC molecules and the substrate surface is also realized and the consequences on orientational behaviour were studied. Both strong and weak surface conditions were used in the investigation of the threshold voltages.

2. Experimental

The measurements were performed on wedge-shaped cells. A SiO double oblique deposition technique (HAUCK) was used to induce surface orientations with different tilt angies ($\varphi_s = 0° - 30°$). A spacer of 18 µm thickness has been arranged at a distance of about 20 mm from the contact place of the two glass plates connected by epoxy resin to build the LC cell. The wedge angle ($\gamma \sim 0.07°$) has been determined from the distance between Newton-rings. Because of deformations of the glass plates caused by the assembling γ was not always constant along the cells. Therefore, γ profils have been maped down.

In our studies we used the nematic basic mixtures SLF 104 (Chisso) and E 5 (BDH). In order to induce a cholesteric phase we added small amounts of S 811 (Merck) which has a helical twisting power of

$$1/c \cdot P_0 = 0.126\ (\mu m \cdot wt.\%)^{-1} \text{ in SLF 104}$$

and

$$1/c \cdot P_0 = 0.117\ (\mu m \cdot wt.\%)^{-1} \text{ in E 5},$$

where c is the concentration of S 811. To realize weak anchoring condition lecithin was added to some mixtures (Tab. 1). The substances were filled into the cells in the cholesteric phase.

Table 1
Composition of the cholesteric mixtures used and their natural pitch P_0

mixture	nematic basic mixture	weight per cent S 811	weight per cent Lecithin	natural pitch P_0/µm
M 1	SLF 104	0.69		12.1
M 2	SLF 104	0.96		7.5
M 3	SLF 104	2.18		3.9
M 4	E 5	0.53		15.5
M 5	E 5	1.07		8.1
M 6	E 5	2.14		4.25
M 4L	E 5	0.53	0.1	15.3
M 5L	E 5	1.07	0.1	8.2
M 6L	E 5	2.14	0.1	4.25

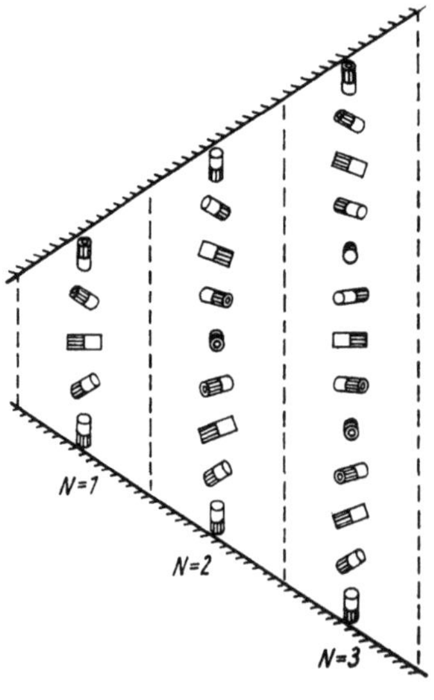

Fig. 1. Orientational pattern in wedge-shaped cells with surface tilt angle $\varphi_s > 0°$

The cells were assembled in a way ensuring that the LC molecules at both surfaces are arranged parallel to each other to get twist angles $\alpha = N \cdot \pi$ (Fig. 1). Beside the thickness variation of the wedge-shaped cells also different P_0 were used to induce various d/P_0 relations (Tab. 1). The measurements were usually performed in the regions of constant twist with $N = 1$ up to $N = 8$.

The optical path difference has been measured with an Ehringhaus rotary compensator in diagonal position between crossed polarizers at the wavelength $\lambda = 551$ nm. Conoscopical methods (Crossland et al.; Komitov et al.) were used to determine the mean tilt angle $\bar{\varphi}$ of the molecules in the cells. Because of the helical twisting the conoscopic pictures became rather blurred. For that reason, we got an accuracy of $\bar{\varphi}$ determination of only about 4—5° at medium and 1—2° at low angles. With the same technique surface tilt angles φ_s were measured in the pure nematic mixtures.

At the contact place of the two glass plates there were no conducting ITO layers. Therefore, in the twist regions with $N = 0$ and, depending on the LC mixture, in a part of those with $N = 1$ there was no possibility to apply an electric field. In the other regions transmission characteristics ($\lambda = 551$ nm) have been recorded up to 15 V at 500 Hz and a sweep rate of 0.05 V · s^{-1}.

3. Results and discussion

3.1. Optical path difference

Some typical results of the measurements of optical path difference Γ between the two eigenwaves of the helical structure are presented in Figures 2 and 3. At the Grandjean lines there are discontinuities in the Γ curves and in the first twisted regions a nonlinear increase with the thickness is distinctly developed. At $\varphi_s = 0°$ the curves in the twisted regions follow that direction given by the straight line of the non-twisted region $N = 0$ (Hauck, Koswig 1985).

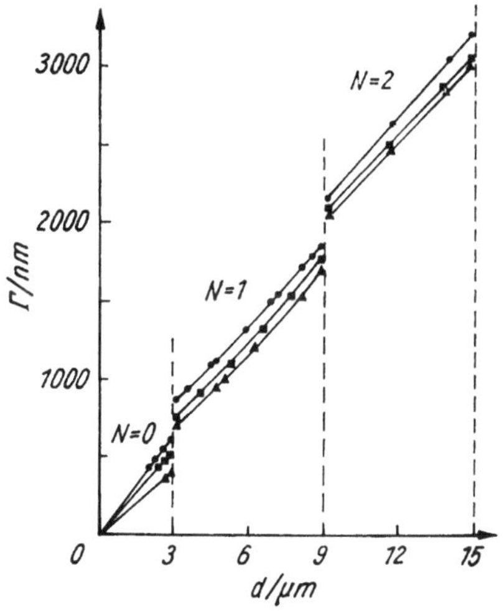

Fig. 2. Optical path difference Γ of mixture M 1 in wedge-shaped cells. The surface tilt angles φ_s are (●) 0°, (■) 7°, and (▲) 20°

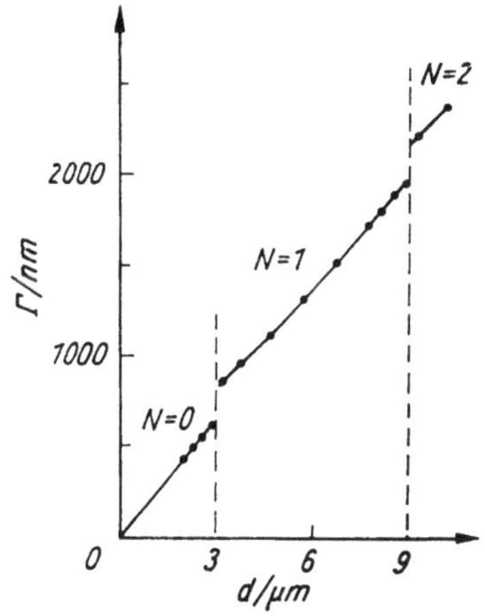

Fig. 3. Optical path difference Γ of mixture M 1 in a cell with surface tilt angle $\varphi_s = 0°$; (solid lines) theoretical curves, and (●) experimental results

This behaviour may be discussed in the following way: By *de Vries* the optical properties of helical systems with $\varphi_s = 0°$ are characterized by

$$\Gamma = (m_2 - m_1)\,\varepsilon_0^{1/2} \cdot d$$
$$m_1 = \{1 + (\lambda')^2 - [4(\lambda')^2 + \beta^2]^{1/2}\}^{1/2} \tag{1}$$
$$m_2 = \{1 + (\lambda')^2 + [4(\lambda')^2 + \beta^2]^{1/2}\}^{1/2}$$

with

$$\varepsilon_0 = \frac{\varepsilon_{\parallel} + \varepsilon_{\perp}}{2},$$

$$\beta = \frac{\varepsilon_{\parallel} - \varepsilon_{\perp}}{\varepsilon_{\parallel} + \varepsilon_{\perp}},$$

$\varepsilon_{\parallel}, \varepsilon_{\perp}$ — dielectric constant parallel and perpendicular to the director

$$\lambda' = \frac{\lambda}{\varepsilon_0^{1/2} \cdot P}$$

P — induced by the surface conditions pitch.

The induced pitch P varies in the twisted regions in the limits

$$\left[\frac{2N-1}{2N} \cdot P_0;\ \frac{2N+1}{2N} \cdot P_0\right].$$

In the centre between the Grandjean lines we have

$$P = P_0 = \frac{2N \pm 0}{2N} P_0,$$

and, in this way,

$$P = \frac{2N + a}{2N} P_0,$$

with

$a = \pm 1$ at the Grandjean lines

$a = 0$ in the centre of the regions.

Assuming a linear change of $P(d)$ in the regions of constant twist, and in this way also a linear change of a, we get

$$a = \frac{d}{d_1} - 2N,$$

with d_1 — thickness at the first Grandjean line ($d_1/P_0 = 0.25$) and

$$P(d) = \frac{d}{2Nd_1} P_0 .$$

N may be expressed by the function of integer part

$$N = INT\left(\frac{d}{2d_1} + \frac{1}{2}\right).$$

Consequently, the consideration of the change in the induced pitch with the thickness transforms λ' to the expression

$$(\lambda')^2 = \frac{\lambda^2}{\varepsilon_0}\left[\frac{1}{2d} \cdot INT\left(\frac{2d}{P_0} + \frac{1}{2}\right)\right]^2. \tag{2}$$

The optical path difference may be calculated from (1) considering (2). Figure 3 shows the good correspondence to the experiment. In the case $\varphi_s = 0°$ the experimental results follow directly from the theory of helical systems.

At tilted boundary orientation there is not such an easy theoretical description. From Figure 2 it is obvious that in contradiction to the tendency in the nontwisted region $N = 0$ Γ shows in the twisted regions only a small decrease in comparison with the planar alignment. This behaviour might be explained by a nonuniform tilt of the molecules in the cell. So, independently of φ_s, the molecules in the bulk of the cell are oriented mainly parallel to the glass substrates (Fig. 1). Because of reasons of symmetry we will assume that in the regions with odd N the tilt angle φ in the centre line between the substrates has to go through $\varphi = 0°$, whereas in the regions with even N this must not be realized to build a continuously tilted structure. This model is in agreement with descriptions for $\alpha = 2\pi$ (BERREMAN) and for $\alpha = \frac{3}{2}\pi$ (SCHEFFER, NEHRING).

Thickness profils can be calculated from the mentioned in Section 2 wedge angle profils. The measurement of Γ in the untwisted region, φ_s and the birefringence of the LC yield the necessary initial values. From the resulting thickness at the Grandjean lines the pitch P_0 can be calculated with a higher accuracy than by the usual method using the distance between the Grandjean lines.

It has been shown that the derived in this way P_0 differ independently from φ_s and the anchoring strength only by $\pm 3\%$ which is within the limits of accuracy of the measurement. Therefore, the structure of the helix in the bulk of the sample is rather independently from the surface conditions.

3.2. Mean tilt angle

The orientation of the molecules in the cells mainly parallel to the surfaces suppose deformation layers near the glass plates where the tilt angle is reduced from the surface quantity to the bulk one. The action of this layer on the optical properties of the whole sample decreases with increasing d. This behaviour should be expressed directly in decreasing mean tilt angle $\bar{\varphi}$ at increasing d (HAUCK, KOSWIG 1985).

In Figure 4 the results of the conoscopical $\bar{\varphi}$ measurement are presented. The predicted behaviour is observable. The decrease of φ in the bulk means that there the forces of helical twisting are stronger than the orientational forces from the surfaces.

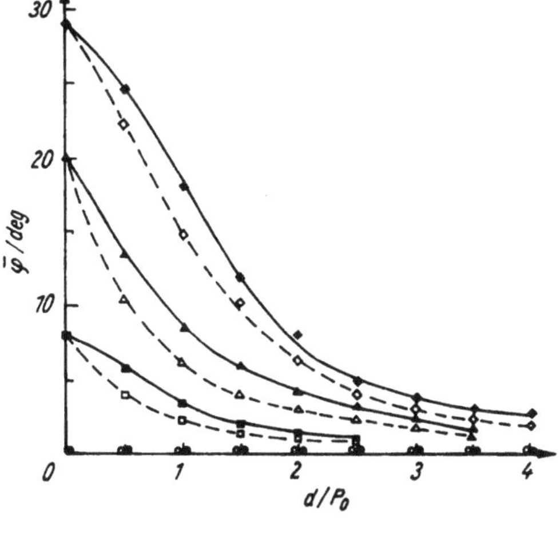

Fig. 4. Mean tilt angle $\bar{\varphi}$ versus ratio between thickness and natural pitch d/P_0: (●) $\varphi_s = 0°$, (■) $\varphi_s = 8°$, (▲) $\varphi_s = 20°$, and (◆) $\varphi_s = 29°$; (solid lines) mixture M 6, and (dashed lines) mixture M 6L

In the case of weak anchoring between the LC molecules at the surface of the cells and the substrate surface, that means in the case where the surface orientation may be influenced by elastical bulk forces, the helical twisting forces cause a further decrease of $\bar{\varphi}$.

No alternation between odd and even N can be detected in Figure 4. Because we measure only the mean tilt, it is impossible to decide if the proposed model of the different, depending on N orientation at the centre line of the cells is correct or not. Unlike that, the general effort of φ to be reduced in the bulk may be used for discussion of the results of electro-optical behaviour. This tendency is also in agreement with the decrease of φ with respect to φ_s in twisted nematic cells ($d/P_0 = 0.25$). In another borderline case (helical structure with $\varphi_s = 90°$, great d/P_0) the relation $\bar{\varphi} \to 0$ is also valid (LIN-HENDEL).

3.3. Threshold voltage

Because at tilted surface alignment there is no true threshold voltage it is convenient to define an apparent threshold voltage (HAUCK, KOSWIG 1984). From the transmission characteristics we defined the voltage of first clear deviation from the zero-field value to be a threshold voltage U_{th}. This corresponds to about 0.5% of the whole transmission change if the accuracy of measurement is taken into consideration.

Figure 5 shows the derived in this way U_{th} (HAUCK, KOSWIG 1985). It has been measured only the values in the centre of the regions of constant twist ($P = P_0$). The connection lines between the measured points have not to correspond to the true values between them. Contrary, there are hints (VERTOGEN, VAN GROESEN; HIRATA et al.) that U_{th} changes between the twist regions in a discontinuous manner. With increased d/P_0 also U_{th} is raised. At $\varphi_s = 0°$ there is no difference between strong and weak anchoring. The helical forces of twisting dominate and U_{th} is determined by the twisted structure in the bulk. With increasing φ_s the threshold voltage of the cholesteric to nematic-homeotropic transition decreases at constant d/P_0. This corresponds to the decrease of apparent thresholds in the nematic LC at $\varphi_s > 0°$. Because at great thickness $\bar{\varphi} \to 0$, the relation U_{th} ($\varphi_s > 0°$) $\to U_{th}$ ($\varphi_s = 0°$) holds as can be seen in Figure 5.

In the case of weak anchoring an increase of U_{th} is developed at the same φ_s and d/P_0, because of the smaller $\bar{\varphi}$. This greater threshold voltage indicates once more the correctness of the developed structural model of tilted helical systems.

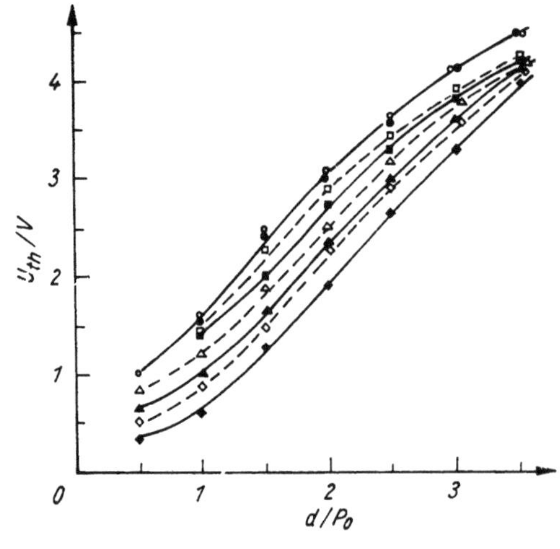

Fig. 5. Threshold voltage U_{th} versus ratio between thickness and natural pitch d/P_0: (●) $\varphi_s = 0°$, (■) $\varphi_s = 8°$, (▲) $\varphi_s = 20°$, and (◆) $\varphi_s = 29°$; (solid lines) mixture M 6, and (dashed lines) mixture M 6L

References

BERREMAN, D. W.: J. Appl. Physics **55**, 806 (1984)
CROSSLAND, W. A., MORRISSY, J. H., NEEDHAM, B.: J. Phys. **D 9**, 2001 (1976)
HAUCK, G.: phys. stat. sol. (a) **62**, K123 (1980)
HAUCK, G., KOSWIG, H. D.: Cryst. Res. Technol. **19**, 1141 (1984)
HAUCK, G., KOSWIG, H. D.: 6th Liquid Crystal Conference of Socialist Countries (Halle, DDR 1985), F 21
HIRATA, S., MATSUZAKI, I., YANGITA, A., TAKO, T.: J. Phys. Soc. Jpn. **50**, 3862 (1981)
KOMITOV, L., HAUCK, G., KOSWIG, H. D.: Cryst. Res. Technol. **19**, 253 (1984)
LIN-HENDEL, C. G.: J. Appl. Physics **53**, 916 (1982)
SCHEFFER, T. J., NEHRING, J.: Appl. Phys. Lett. **45**, 1021 (1984)
TOGASHI, S., SUZUKI, T., AOYAMA, T., SHIMIZU, H.: Mol. Cryst. Liq. Cryst. **69**, 117 (1981)
VERTOGEN, G., VAN GROESEN, E. W. C.: J. Chem. Phys. **76**, 2043 (1982)
DE VRIES, H.: Acta Cryst. **4**, 219 (1951)

(Received March 17, 1987)

Authors' address:

Dr. G. HAUCK, Prof. H. D. KOSWIG
Zentralinstitut für Elektronenphysik der Akademie der Wissenschaften der DDR
Hausvogteiplatz 5/7
Berlin
DDR-1086

Book Review

H.-O. Peitgen, P. H. Richter

The Beauty of Fractals

Springer-Verlag Berlin-Heidelberg-New York-Tokyo 1986. 184 figs., many in color. XII + 199 pp. Hard cover DM 78.–, ISBN 3-540-15851-0

The concepts of self-similarity and scale invariance have arisen independently in the areas of phase transitions and fractal geometry. Now it is well accepted that fractals are all around us and the concept of fractals has extended to quite different fields of science. One convinaing argument in favour of fractals is their beauty. The advent of computer graphics facilities encouraged several authors to visualize the mathematical concept of fractals. The authors of the present book are mathematicians and physicists working in the field of chaotic dynamics. The positive response that their first pictures of fractals found in exhibitions led them to the idea of publishing the most beautiful examples. The volume is divided into 10 sections, each starting with a special mathematical problem. The headings are: Verhulst Dynamics, Julia Sets, Sullivan's Classification of Critical Points, Mandelbrot Set, Hubbard Trees, Cayley's Problem, Newton's Method for Real Equations, A Discrete Volterra-Lotka System, Yang-Lee Zeros, Renormalization. So, the book is primarily a treatise of the mathematics and physics of special phenomena.

However, readers not interested in maths will find it exciting to look simply at the beautiful pictures included. Those who want to redo them on their own computers get a few hints by the authors at the end of the book.

Moreover four experts in the field (B. B. Mandelbrot, A. Douady, G. Eilenberger, H. W. Franke) contributed their own view to this new development which make the book even more attractive.

The good quality of the present edition may serve as strong argument for non-mathematically minded readers to buy it. One can be sure that those who are fascinated by the idea of connecting art and science will enjoy this publication. P. Paufler

Cryst. Res. Technol.	22	1987	10	K159–K162

Short Notes

E. MORHÁČOVÁ

Institute of Physics of the Electro-Physical Research Centre, Slovak Academy of Sciences, Bratislava, Czechoslovakia

The Average Grain Size of Calcia-stabilized Zirconia Doped with Copper and Iron

1. Introduction

The properties of ceramic materials are strongly dependent on their microstructural parameters (UNDERWOOD et al.). Average grain size is a very important parameter used in determination of the electrical and mechanical properties of these materials. It is known that the electrical conductivity of polycrystalline materials at low temperatures is determined in essence by the grain boundary conductivity. For the determination of this conductivity contribution we need to know the average grain size.

2. Grain size determination

The polycrystalline ceramics have a grained polyhedral structure (SALTYKOV). Since it is not oriented, the three-dimensional view of the sample can be replaced by a two-dimensional one observed under the light microscope on the ground and polished surface.

Calcia-stabilized zirconia samples doped with iron and copper were investigated (HARTMANOVÁ et al.). The surface of the pellets was ground and polished. To make the outlines of the grains more visible, the grain boundaries were etched by a mixture of $HNO_3 + HF + HCl$ till the grains were clearly visible (from 2 to 15 minutes). The micrographs were prepared with the Superprobe JCXA-733. The pictures of the sample were redrawn on transparent foil.

For the semiautomatical evaluation of microstructural parameters the program MAPOC was developed (MORHÁČOVÁ). This program allows to redraw a microstructure picture from transparent foil on the TV screen and simultaneously record into the computer memory (Sinclair Spectrum ZX). After the black and white picture being in the memory matrix with dimensions 254 · 174, the program uses the intercept length method to determine the average grain size (CASE et al.). The regular grid is drawn across the picture in computer memory and all intercepts are registered. At the same time the experimental counts are divided into 15 columns of a histogram. The average grain size is calculated subsequently using formulae:

$$\bar{L} = L/N , \tag{1}$$

$$\bar{D} = C \cdot M \cdot \bar{L} , \tag{2}$$

where $\bar{L}$ is the average intercept length, L is the sum of the lengths of all intercepts, N is the number of all found intercepts, $\bar{D}$ is the final average grain size, C is the constant that transforms the two-dimensional picture of a polished surface to the three-dimensional real shape of a grain (Mendelson) and M is the magnification.

All samples were evaluated using this program. The results are given in Table 1 and on the Figures 1—7.

Table 1

Average grain sizes of calcia stabilized zirconia samples in dependence on dopant concentration

Sample	$\bar{D}/\mu m$
$(ZrO_2)_{0.85}(CaO)_{0.15}$	47.45
$(ZrO_2)_{0.85}(CaO)_{0.15}(Fe_2O_3)_{0.005}$	2.10
$(ZrO_2)_{0.85}(CaO)_{0.15}(CuO)_{0.01}$	6.42
$(ZrO_2)_{0.857}(CaO)_{0.12}(CuO)_{0.02}(Fe_2O_3)_{0.003}$	10.50
$(ZrO_2)_{0.85}(CaO)_{0.15}(Fe_2O_3)_{0.03}$	7.98
$(ZrO_2)_{0.85}(CaO)_{0.15}(CuO)_{0.05}$	10.07

3. Results

The average grain sizes were calculated for the calcia-stabilized zirconia with various concentrations of dopants. Dependence of this parameter on the addition concentration is evident from Figure 1. Adding the small concentration of the dopant

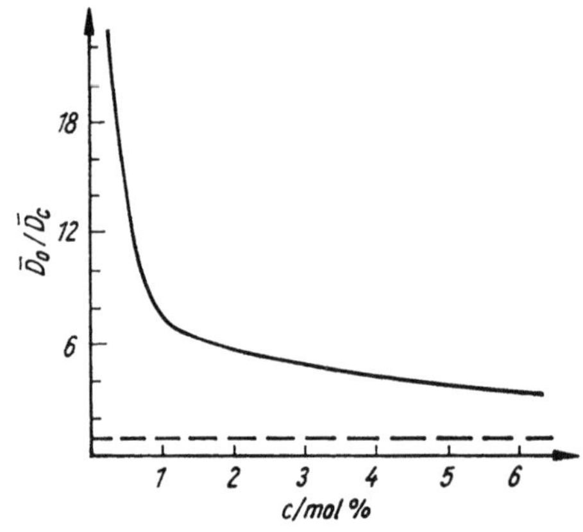

Fig. 1. Dependence of average grain size on the dopant concentration. $\bar{D}_0$ – average grain size of the sample without dopant, $\bar{D}_c$ – average grain size of the sample with aopant concentration c

(Fe_2O_3 — 0.5 mol.%) causes rapid decreasing of the average grain size with regard to grain size material without dopant (2.10 μm — 47.45 μm). The average grain size grows again with adding of higher concentrations of the dopants (see Tab. 1). Minimum concentration of the dopant (0.5 mol.%) causes decomposition of solid solution. Particles of Fe_2O_3 are continuously dispersed by diffusive process and grain growth is restricted. Increasing the concentration of the dopant causes that particles of the dopant begin to gather into the precipitates on grain boundaries and grains can grow again.

The experimental counts plotted in histograms (Figs. 2—7) show clearly that zirconia ceramics have lognormal grain size distribution. Pure calcia-stabilized zirconia

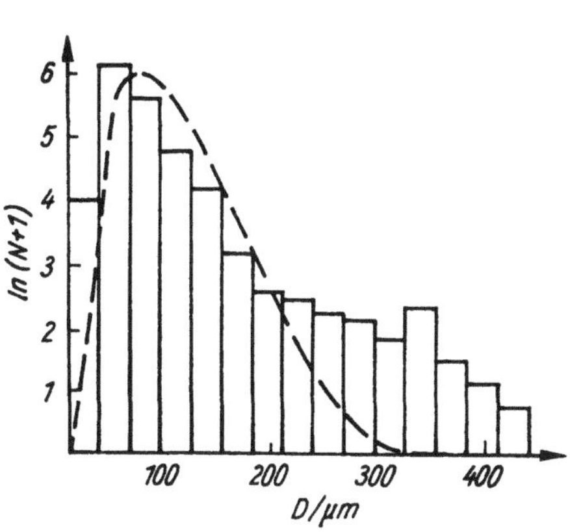

Fig. 2. Histogram of (ZrO_2) (CaO) sample

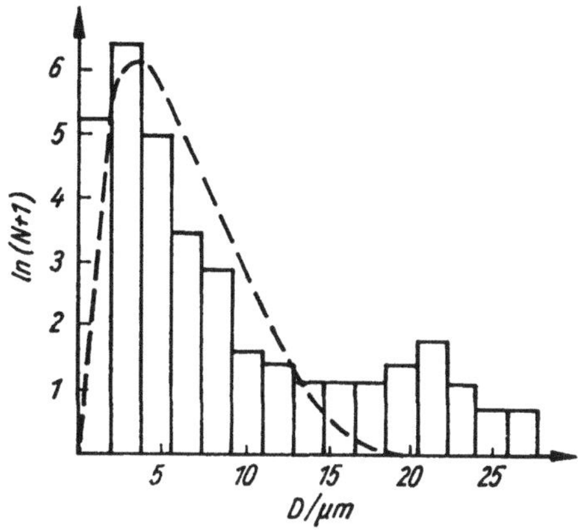

Fig. 3. Histogram of (ZrO_2) (CaO) $(Fe_2O_3)_{0.005}$ sample

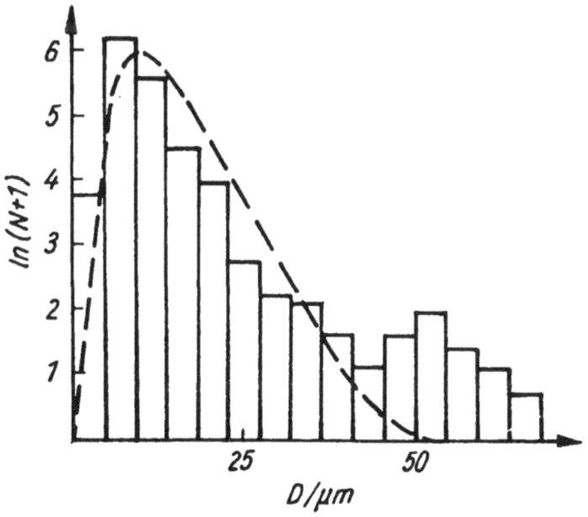

Fig. 4. Histogram of (ZrO_2) (CaO) $(CuO)_{0.01}$ sample

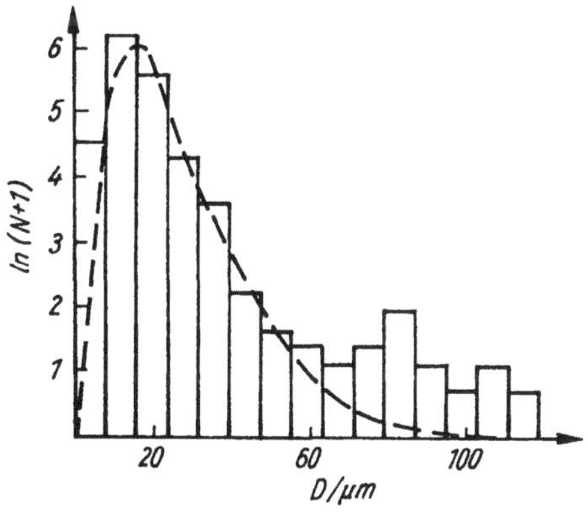

Fig. 5. Histogram of (ZrO_2) (CaO) $(CuO)_{0.02}$ $(Fe_2O_3)_{0.003}$ sample

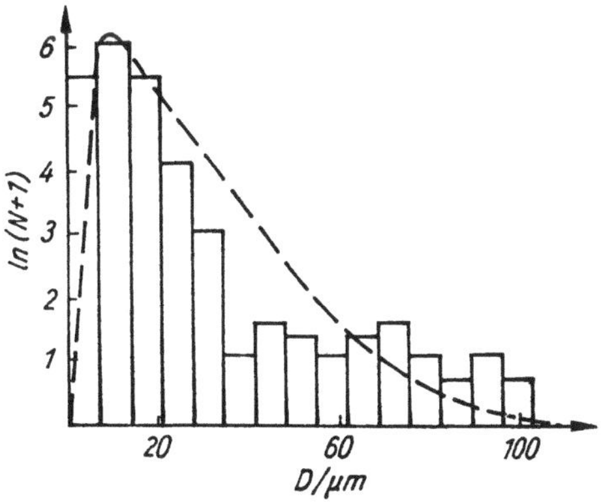

Fig. 6. Histogram of (ZrO_2) (CaO) $(Fe_2O_3)_{0.03}$ sample

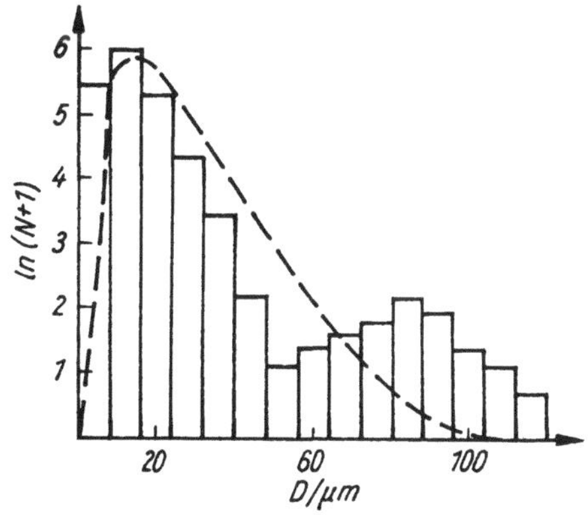

Fig. 7. Histogram of (ZrO_2) (CaO) $(CuO)_{0.05}$ sample

has only one peak in the distribution (Fig. 2). By adding of the dopant another small peak in the last thirdth of histogram becomes to be visible (Figs. 3—4). This peak moves to the second half of the histogram at the concentration of the dopant of 5 mol % (Fig. 7). This is the consequence of repeated grain growth permitted by the precipitation mentioned above.

4. Conclusions

Measured results showed that there are not only powder characterization and sintering conditions that influence the grain evolution. Also purity of material and dopant concentration are remarkable factors of grain growth. This work attempts to outline the extent of dependence of the microstructure parameters on the dopant concentrations.

References

CASE, E. D., SMYTH, J. R., MONTHEI, V.: J. Amer. Cream. Soc. **64**, C-24 (1981)

HARTMANOVÁ, M., MACHOVIČ, L., KOLLER, A., HANIC, F., MIŠÁNIK, B.: Solid State Ionics **14**, 93 (1984)

MENDELSON, M.: J. Amer. Ceram. Soc. **52**, 443 (1969)

MORHÁČOVÁ, E.: MAPOC — The program for microstructural analysis of polycrystalline ceramics, Internal report, Bratislava 1986

SALTYKOV, S. A.: Stereometricheskaya metallografiya, Moscow 1970

UNDERWOOD, E. E., COLCORD, A. R., WAUGH, R. C.: Ceramic Microstructures, New York 1968

(Received April 7, 1987)

Author's address:
Eva MORHÁČOVÁ
Institute of Physics, Electro-Physical Centre
Slovak Academy of Sciences
Dúbravská cesta 9
842 28 Bratislava
Czechoslovakia

Cryst. Res. Technol.	22	1987	10	K163–K165

U. KRETZSCHMAR, F. NEISSENDORFER

Zentralinstitut für physikalische Chemie der Akademie der Wissenschaften der DDR, Berlin

RDF Studies of the Short Range Order of Mechanically Treated MgO Powder

1. Introduction

The mechanical treatment of powder in mills may change the structure of the solid. In earlier publications (BERNHARDT, HEEGN; STEINIKE et al.) it has been shown that by a mechanical treatment (vibration mill) preferably a disturbance of the surface-near range of the grains of the powder is attained (high dislocation density, KRETZSCHMAR et al.). In the present investigation we aim at determining the short order range of mechanical distortet substances by mean of X-ray wide-angle diffraction. We discuss the change of short order range from calculation of difference radial distribution functions (DRDF).

2. Experimental

Initial substance: MgO powder (spezific area 1.4 m^2/g)
Mill: Vibration mill-typ LS 60 (VEB KEFAMA Katzhütte)
Planetary mill-typ EU 2 × 150
X-ray scattering measurements of the mechanically induced defects were carried out on a diffractometer (HZG3 VEB Freiberger Präzisionsmechanik). MoK_α radiation was diffracted at the sample and then monochromatized by reflection on an LiF crystal. Every sample was measured 6 times with equidistant steps of 0.1. The observed s-range was $6\ nm^{-1} < s < 120\ nm^{-1}$ ($s = 4\pi \sin \vartheta/\lambda$). The measured intensity was corrected (background, scattering area, absorption, polarisation, incoherent radiation) and then scaled to absolute intensities (HAJDU; STEEB). The data analysis was carried out with a BESM 6 computer (ZfR der AdW). The used program for the calculation of DRDF was described by NEISSENDORFER in detail.

3. Results and discussion

The difference-radial distribution functions (DRDF) are shown in Figures 1 and 2. The function was calculated according to

$$\mathrm{DRDF}(r) = \frac{2r}{\pi} \int\limits_{s_{\min}}^{s_{\max}} s i(s) \sin(sr)\, ds ,$$

where r is the interratomic distance and $i(s)$ is the experimental structure function (STEEB).

The curves (Figs 1 and 2) appear first very similar. They are characterized by two peaks in the range $r < 0.35$ nm. The first peak of the DRDF at around 0.210 nm

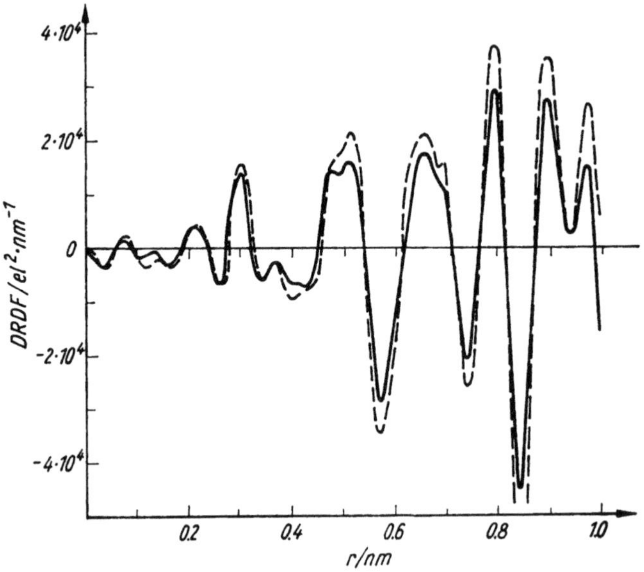

Fig. 1. DRFD of treated MgO: initial substance (dashed line); vibration mill 64 hours (solid line)

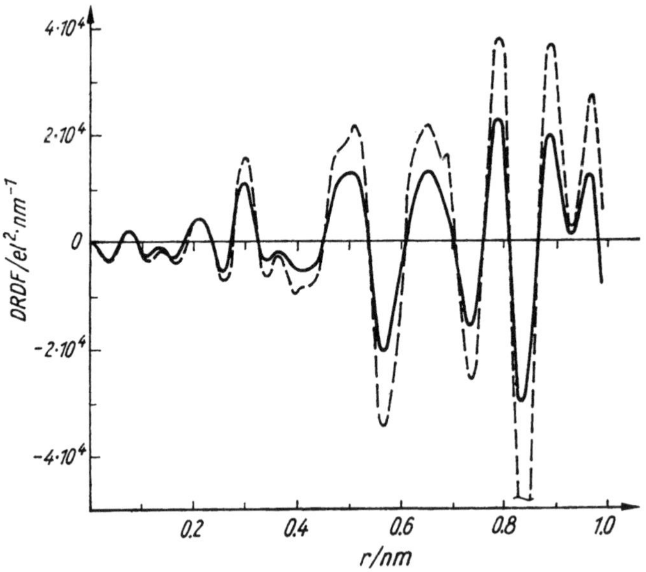

Fig. 2. DRDF of treated MgO: initial substance (dashed line); planetary mill 150 min (solid line)

corresponded to the Mg–O distance in accord with the literature (HELLWEGE). The second peak appeared at around 0.298 nm which corresponded to both the distances between Mg–Mg and the O–O distance in first coordination (Tab. 1).

The area of the second peak changes. In the wide range r about 0.5 nm the height of the peak's decrease corresponds to the intensity of the mechanical treatment. Therefore in our case the lattice is evidently distorted.

Table 1
Interatomic distance ± 0.003 nm

atomic arrangement	calculated from lattice parameters	initial substance	vibration mill		planetary mill 150 min
			32 h	64 h	
O–Mg	0.211	0.210	0.213	0.221	0.205
O–O/Mg–Mg	0.298	0.300	0.297	0.298	0.295
O–Mg	0.365	0.365	0.365	0.363	0.365
O–O/Mg–Mg	0.421	—	—	—	—
O–Mg	0.471	—	—	—	—
O–O/Mg–Mg	0.516	0.510	0.510	0.515	0.510

4. Conclusion

The interpretation of DRDF obtained by X-ray diffraction measurements allows us to conclude that a distortion of the unit cell is achieved only by means of intensive mechanical treatment. The surface-near ranges of the grains which contain a high concentration of dislocations may be regarded as a kind of "distorted" phase which is connected with a distortion of the unit cell.

The authors wish to thank Mr. H.-P. Hennig for the mechanical treatment of powders in the vibration mill. The treatment in the planetary mill was carried out in the Institute for Solid State Chemistry, Novosibirsk.

References

Bernhardt, C., Heegn, H.: 4. Europ. Symp. Zerkleinern, Nürnberg 1975
Hajdu, F.: Acta Cryst. A28, 250 (1972)
Hellwege, K. H.: Einführung in die Festkörperphysik, Berlin-Heidelberg 1976
Kretzschmar, U., Ebert, I., Steinike, U., Hennig, H.-P.: Cryst. Res. Technol. 17, 257 (1982)
Neissendorfer, F.: in preparation
Steeb, S.: Springer Tracts of Modern Physics 47, 1 (1968)
Steinike, U., Ebert, I., Jedamzik, J., Hennig, H.-P., Kretzschmar, U.: Kristall u. Technik 13, 1229 (1978)

(Received April 10, 1987)

Authors' address:

Dr. U. Kretzschmar, Dipl. Phys. F. Neissendorfer
Zentralinstitut für physikalische Chemie der Akademie der Wissenschaften der DDR
Rudower Chaussee 5
DDR-1199 Berlin

Cryst. Res. Technol.	22	1987	10	K166–K168

Department of Physics, Sardar Patel University, Vallabh Vidyanagar, Gubjarat, India

C. C. DESAI, M. S. V. RAMANA

Some Aspects of the Magnetic Susceptibility of Ferroelectric LHP and LNP Crystals

1. Introduction

The study of magnetic susceptibility at different magnetic field is an important tool to probe into the magnetic properties of the materials. Considerable amount of work along these lines has been carried out on a variety of materials, for example, scheelite compounds (GODBOLE), and tin iodates (DESAI, RAI) yielding valuable information. The information on magnetic properties of ferroelectric crystals such as LHP (Lead Hydrogen Phosphate) and LNP (Lead Nitrate Phosphate) are meagre. In this paper, we report magnetic susceptibilities of these crystals using Faraday Gauy balance technique.

LHP and LNP crystals are of considerable interest, particularly in basic studies of their phase transitions and of their electronic, dielectric, ferroelectric, optical and other pertinent characteristics. These crystals are used in transducers and in many linear and non-linear mechanical devices. They are also suitable for testing microscopic theory of ferroelectricity. An attempt has been made to understand the magnetic properties of these materials.

2. Theory

The magnetic permeability of a compound in a magnetic field is given by the relation:

$$\mu = 1 + 4\pi k \,, \tag{2.1}$$

where k is the volume susceptibility of the material. The quantity often calculated is the magnetic susceptibility per gram (χ_g) which is defined as:

$$\chi_g = \frac{k}{d} \,, \tag{2.2}$$

where d is the mass density of the material used for study.

Diamagnetism arises from the electrons in closed shells tending to orient their planes of rotation so as to suffer minimum interaction with the imposed magnetic field. All the diamagnetic substances are characterised by negative values of k and χ_g. Paramagnetism arises from the angular moments, orbital and/or spin, associated with the unpaired electrons. For all paramagnetic materials k and χ_g are positive. Ferromagnetism and antiferromagnetism are sub-divisions of the paramagnetism and are known as "co-operative phenomena", which means that they arise when the paramagnetic centers within a sample intersect magnetically with each other.

3. Experimental procedure

Single crystals of LHP and LNP were grown using the controlled reaction between lead nitrate and orthophosphoric acid solutions. The crystals were examined by scanning electron microscope, EDAX, density measurements and thermogravimetric analysis, these techniques confirmed that they were LHP (monoclinic, space group P 2/a) and LNP (monoclinic, space group P 2_1/c).

The measurements of magnetic susceptibility of LHP and LNP samples were carried out using Faraday method (BATES). The finely powdered crystals of LHP and LNP were packed in a small container of known magnetic susceptibility. This container was hung between two powerful electromagnetic pole pieces of a Gauy balance, the sensitivity of the balance was 10^{-5} g. Initially samples were weighed accurately in the absence of the magnetic field. The magnetic field was increased by increasing the current from 4 to 8 A through the coils of electromagnets. The magnetic field strength was measured with a Gauss-meter. The change in weight, i.e. reduced mass of the samples, after applying the magnetic field was measured. The positive gram susceptibility, χ_g, characteristic of a paramagnetic material was calculated using the formula:

$$\chi_g = \frac{\alpha + \beta\, dW}{W}\,, \tag{3.3}$$

where

α is a constant which represents the correction factor due to the displacement of air,

β is the constant for the glass container at a given current,

W is the weight of the sample, and

dW is the charge in weight when a magnetic field is applied, i.e. $\Delta W - \delta$ is the apparent change in weight (in mg) of the tube containing W g of the sample on the application of the field and δ is the difference in weight of the empty tube on the application of the field.

The gram susceptibility of the material was multiplied by its molecular mass to obtain molar susceptibility (χ_M).

4. Results and discussion

Crystals of LHP and LNP are usually transparent and white in colour and rectangular in shape. These crystals were finely ground and the resulting powder (particle size 150 µm) were packed in a Gauy tube of known magnetic susceptibility. The experiments of magnetic susceptibility measurements were repeated five times and the average change in weight was computed for a given current. The accuracy in the measurement of reduced mass is $\pm 5\%$. The readings of the Gauy balance were recorded when the values became steady. The values were quite consistent and reproducible. The magnetic susceptibilities of LHP and LNP samples obtained at room temperature (28 °C) are recorded in Table 4.1.

From the Table 4.1 it is clear that:

1. due to the application of the magnetic field the reduced weights of all samples is increased,
2. the magnetic susceptibilities (χ_g and χ_M) are independent on the current, and
3. χ_M of LNP samples are greater than those of LHP samples, and
4. the positive values of the susceptibilities thereby suggest that LHP and LNP are paramagnetic materials. This may be attributed to the presence of lead (unfilled p shells) in them.

The ferromagnetism is usually a sub-division of the paramagnetism and is known as "co-operative phenomenon" which thereby initiate when the paramagnetic centers within materials intersecting magnetically with each other. In fact, para-

Table 4.1
Magnetic susceptibilities of LHP and LNP

Particulars of the sample	current in A	loss in weight ΔW in g	$\alpha \times 10^{-6}$	$\beta \times 10^{-6}$	gram susceptibility χ_g ($\times 10^{-9}$ e.m.u.)	molar susceptibility χ_M ($\times 10^{-9}$ e.m.u.)
LHP	3	0.00385	0.02949	0.8026	7.5654	2.2939
	4	0.00411	0.02949	0.7897	7.5665	2.2939
	5	0.00532	0.02949	0.6339	7.5683	2.2944
	6	0.00651	0.02949	0.4861	7.5691	2.2947
	7	0.00723	0.02949	0.4176	7.5689	2.2946
	mean value	—	—	—	7.5676	2.2942
LNP	3	0.00401	0.02949	0.8026	7.4114	8.7364
	4	0.00438	0.02949	0.7897	7.4171	8.7361
	5	0.00548	0.02949	0.6379	4.4113	8.7363
	6	0.00657	0.02949	0.4861	7.4115	8.7365
	7	0.00696	0.02949	0.4176	7.4172	8.7362
	mean value	—	—	—	7.4113	8.7363

magnetic characteristics generates from: (i) the angular momentum and (ii) orbital spins associated with the un-paired electrons.

That the positive values of molar susceptibilities of ferroelectric LHP and LNP samples catagorise them to be paramagnetic materials.

5. Conclusions

1. The gram- and molar susceptibilities of ferroelectric LHP and LNP crystals have been determined.
2. The gram- and molar susceptibilities are independent on the magnetic field.
3. Molar susceptibility of LNP is greater than that of LHP crystals.
4. The positive values show that LHP and LNP are paramagnetic materials.

The authors are thankful to Council of Scientific and Industrial Research (CSIR), New Delhi for the sanction of the project, entitled "Growth and characterization of Ferroelectric crystals". One of us (MSVR) is very grateful to Dr. M. Rama Rao and Mrs Anandaraman (Hyderabad) for their keen interest in the present work.

References

Bates, L. F.: Modern Magnetism, Cambridge 1951 (U. K.)
Desai, C. C., Rai, J. L.: J. Mater. Sci. Letters 2, 467 (1983)
Godbole, R. S.: Ph. D. Thesis, Sardar Patel University, India

(Received April 27, 1987)

Authors' address:

Professor C. C. Desai, Mr. M. S. V. Ramana
Department of Physics, Sardar Patel University
Vallabh Vidyanagar 388120
Gujarat state, India

Cryst. Res. Technol.	22	1987	10	K169–K171

B. VENGATESAN, N. KANNIAH, P. RAMASAMY

Crystal Growth Centre, Anna University, Madras, India, and
Government Arts College, Villupuram, India

Growth of Cadmium Sulphide Thin Film by Open Tube Chemical Vapour Transport Method

1. Introduction

Even though single crystals of CdS have been grown by a variety of techniques (SHARMA, MALHOTRA; MOULIN et al.; BOHAC, GAUMANN), the sublimation techniques have been widely used for this purpose. In sublimation techniques, ambients are either vacuum or inert gas or an elemental (Cd or S) component atmosphere. A variety of temperatures and temperature profiles are described usually ranging from 1100 °C to 1250 °C; lower values are reported with seeding procedures (DIERSEN, GABOR) or under particular vacuum preheating conditions of the charge (SHARMA, MALHOTRA). The disadvantages of these sublimation methods are (i) the high temperature involved, close to the softening of quartz and (ii) the high sensitivity to growth rate, stoichiometric deviations and impurities in the charge (FAKTOR, GARRETT). Thus large changes in growth rate and even suppression of growth are often reported under the same growth conditions.

Using H_2–CdS vapour growth technique MOULIN et al., and ITO, OHSAWA obtained epitaxial CdS layers on GaAs and InP substrates although polycrystalline layers were deposited at substrate temperatures 540 °C and 500 °C, respectively.

Formation of CdS thin film on glass substrate by Open Tube Chemical Vapour Transport method has been reported in this investigation. Iodine is employed as the transporting agent. The source temperature and the substrate temperature have been maintained at much lower values than those reported in literature (MOULIN et al.; ITO, OHSAWA).

2. Experimental

Analar grade CdS (5 grams) was mixed with the transporting agent iodine at the concentration of 5 mg/cm³. This mixture was taken in a porcelain boat with a diameter of 0.5 cm and a length of 3 cm. The boat was placed in an open quartz tube with a diameter of 5 cm and a length of 50 cm. This quartz tube was placed in a two zone temperature furnace as shown in the Figure 1. The three glass slides with a breadth of 1.5 cm and a length of 5 cm each were well cleaned first with hot distilled water and then cleaned with Aqua Regia and left for 6 hours. Then they were repeatedly cleaned with triple distilled water. The three glass slides were arranged inside the tube. The source material was in Zone I and the slides were in Zone II.

At first the glass slides were maintained for 24 hours at a higher temperature than the source material to avoid the spurious nucleation. Then the temperature of this Zone II was gradually lowered to 350 °C and the temperature of Zone I was gradually increased to the maximum temperature of 850 °C.

Due to the temperature profile of the furnace, the three slides were maintained at different temperatures. The experiment was carried out for a period of 6 hours. Thin film of CdS was formed uniformly over the glass slides.

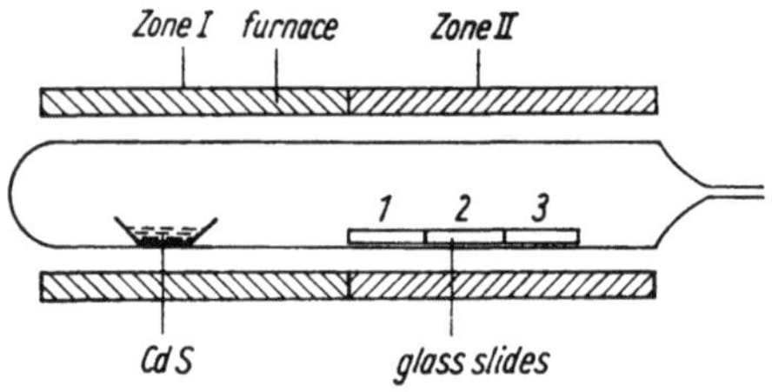

Fig. 1. Schematic diagram of the vapour growth apparatus

3. Results and discussion

The formation of CdS thin film can be explained by considering the following transport reaction

$$CdS_{(s)} + I_{2(g)} \rightleftharpoons CdI_{2(g)} + 1/2\ S_{2(g)}\ .$$

The less volatile CdS reacts with iodine to form a highly volatile CdI_2 vapour which diffuses to the colder zone due to the temperature gradient inside the tube. At the colder zone CdI_2 and S_2 react to form $CdS_{(s)}$ due to the shift of the equilibrium with the decrease in temperature. The backward reaction is feasible due to the negative free energy change at the lower temperature. Thus CdS is deposited on the glass substrates.

The thin films were analysed using Leitz Wetzlar Microscope at a magnification of 500. When the temperature was 350 °C, a hexagonal platelet layer is deposited (Fig. 2). The patterns are similar to ice crystal structure. Similar patterns are also

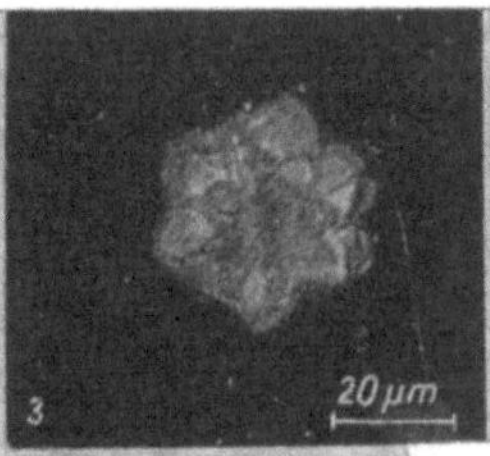

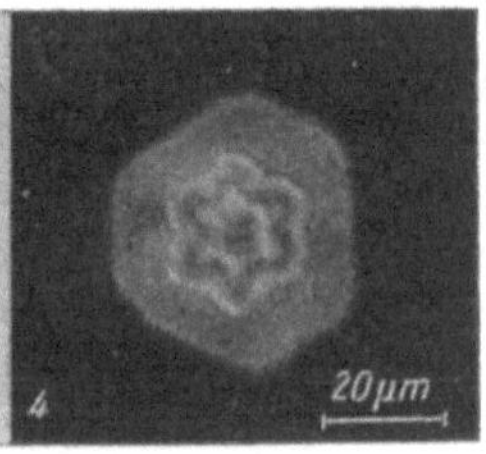

Fig. 2. Hexagonal platelet layer of CdS

Fig. 3. Central flower-like pattern of CdS

Fig. 4. Hexagonal platelet formation with central flower-like pattern

Fig. 5. Grain boundary line of CdS

obtained in the case of PbI_2 and AgI. In Figure 2, some flower like patterns are also observed as in the case of natural ice crystals. The growth pattern can be explained as follows. At first the central flower-like pattern is formed (Fig. 3). Then the gaps in between the petals are packed with the same material loosely and thereby a complete hexagonal formation is obtained (Fig. 4). The growth habit of these types of crystals is generally a layered structure. Figure 5 shows the grain boundary line due to the change in the temperature in slide 2. This is further confirmed by the third slide which shows the dendritic pattern (Fig. 6). The dendritic CdS crystal has been formed due to higher temperature gradient.

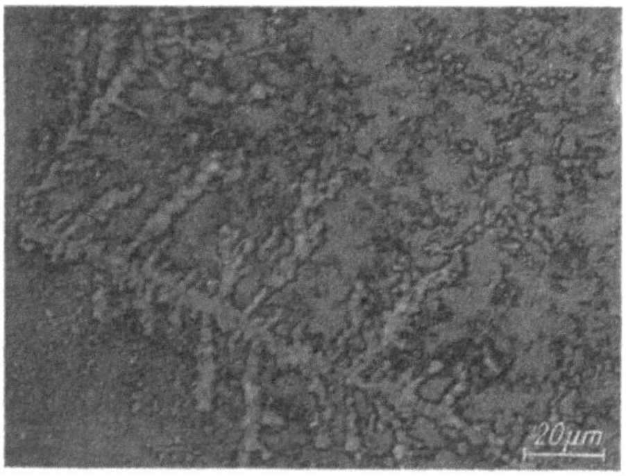

Fig. 6. Dendritic pattern of CdS

4. Conclusion

Contrary to the existing deposition temperature of 500 °C and 540 °C when hydrogen was used as a transport agent for deposition of CdS on InP and GaAs, respectively, in the present study the deposition temperature of CdS has been observed at a much lower temperature of 350 °C with iodine as a transporting agent. The experimental results explain the growth behaviour of CdS.

One of the authors (B. V.) is thankful to C.S.I.R. for the award of a Junior Research Fellowship. The authors are grateful to Dr. S. Vasudevan for his assistance in the preparation of photos.

References

Bohac, P., Gaumann, A.: J. Cryst. Growth **26**, 171 (1974)
Diersen, G. H., Gabor, T.: J. Cryst. Growth **43**, 572 (1978)
Faktor, M. M., Garrett, I.: Growth of crystals from the vapour, London 1974
Ito, K., Ohsawa, T.: Jpn. J. Appl. Physics **16**, 11 (1977)
Moulin, M., Huber, A., Pugue, M.: J. Cryst. Growth **17**, 212 (1972)
Sharma, S. D., Malhotra, L. K.: J. Cryst. Growth **10**, 199 (1971)

(Received May 13, 1987)

Authors' addresses:

B. Vengatesan*), N. Kanniah**), and P. Ramasamy*)

*) Crystal Growth Centre, Anna University
Madras-600025, India

**) Government Arts College
Villupuram-605602, India

Crystal Research and Technology

Zeitschrift für experimentelle und technische Kristallographie

Edited by H. Neels, Leipzig

Papers in Number 11, 1987

Review

Original Papers

AKADEMIE-VERLAG · BERLIN

Continued from cover page 4

Short Notes

Book Reviews

Crystal Research and Technology
Volume 22 1987 Number 10

Contents

Original Papers

Continued on cover page 3

Crystal Research and Technology is indexed in Current Contents/PC & ES.

Bei Fragen zur Produktsicherheit wenden Sie sich bitte an:
If you have any questions regarding product safety,
please contact:

Walter de Gruyter GmbH
Genthiner Straße 13
10785 Berlin
productsafety@degruyterbrill.com